SpringerBriefs in Energy

For further volumes:
http://www.springer.com/series/8903

Erik Kjeang

Microfluidic Fuel Cells and Batteries

 Springer

Erik Kjeang
Fuel Cell Research Laboratory
School of Mechatronic Systems Engineering
Simon Fraser University
Surrey, BC, Canada

ISSN 2191-5520   ISSN 2191-5539 (electronic)
ISBN 978-3-319-06345-4   ISBN 978-3-319-06346-1 (eBook)
DOI 10.1007/978-3-319-06346-1
Springer Cham Heidelberg New York Dordrecht London

Library of Congress Control Number: 2014939516

Printed on acid-free paper

Springer is part of Springer Science+Business Media (www.springer.com)

For Gunilla

Preface

Summarizing the initial 10 years of research and development in the field of microfluidic fuel cell and battery technology for electrochemical energy conversion and storage, this SpringerBrief is the first book dedicated to this emerging field. Written at a critical juncture, where strategically applied research is urgently required to seize impending technology opportunities for commercial, analytical, and educational utility, this book is a comprehensive resource for current and prospective researchers in the general area of membraneless, microfluidic electrochemical energy conversion.

I have been active in this field since 2005, when I joined the University of Victoria as a Ph.D. student, inspired by the notion of a low-cost fuel cell without membrane or catalyst. My initial research endeavors culminated in a Ph.D. dissertation entitled "Microfluidic fuel cells" [1], which was subsequently awarded with the Governor General's Gold Medal and launched my career as a researcher and scholar. Since 2009, I have continued my research on microfluidic fuel cells as a faculty member at Simon Fraser University (SFU), where I established the SFU Fuel Cell Research Laboratory (FCReL) and expanded the scope of this research to include microfluidic redox flow batteries.

Since the goal of the book is to provide a comprehensive resource for both research and technology developments, it features extensive descriptions of the underlying fundamental theory, fabrication methods, and cell design principles, as well as a thorough review of previous contributions in this field and a concluding chapter with recommendations for further work. It builds substantially on information collected over the last 10 years and draws specifically on our previously published review articles in *Journal of Power Sources* [2] and *Biomicrofluidics* [3] and book chapter in *Micro Fuel Cells: Principles and Applications* [4], as well as a recent review manuscript on co-laminar flow cells for electrochemical energy conversion [5]. It is hoped that this book will enable new research groups to develop the next generation of microfluidic electrochemical cells.

I wish to express my sincere gratitude to everyone who has contributed either directly or indirectly to the publication of this book. In particular, I wish to thank my former supervisors at the University of Victoria, Dr. David Sinton, Dr. Ned Djilali, and Dr. David Harrington. I thank all current and past SFU FCReL students who have participated in our microfluidic fuel cell and battery projects, especially Marc-Antoni Goulet and Dr. Jin Wook Lee who have had leading roles over the past several years and generated major advances in this field and including, but not limited to, Bernard Ho, Deepak Krishnamurthy, Erik Johansson, Xiaoye Liang, Jun Hong, Omar Ibrahim, Aronne Habisch, Dean Chen, Spencer Arbour, Jeetinder Ghataurah, Chris de Torres, Stephan Rayner, Jeffrey To, Nader Moradi, David Afonso, Dan Latuszek, Larry Hoang, Peter Hsiao, Sandeep Sanghera, and Christopher Stewart.

This research was supported by the Natural Sciences and Engineering Research Council of Canada, Western Economic Diversification Canada, Canada Foundation for Innovation, British Columbia Knowledge Development Fund, and Simon Fraser University.

Surrey, BC, Canada   Erik Kjeang

References

1. E. Kjeang, *Microfluidic Fuel Cells*, PhD Dissertation, University of Victoria, 2007
2. E. Kjeang, N. Djilali, D. Sinton, Microfluidic fuel cells: a review. J. Power Sources. **186**, 353–369 (2009)
3. J.W. Lee, E. Kjeang, A perspective on microfluidic biofuel cells. Biomicrofluidics. **4**, 041301 (2010)
4. E. Kjeang, N. Djilali, D. Sinton, Advances in microfluidic fuel cells, in *Micro Fuel Cells: Principles and Applications*, ed. by T.S. Zhao (Elsevier B.V., 2009). ISBN: 978-0-12-374713-6
5. M.A. Goulet, E. Kjeang, Co-laminar flow cells for electrochemical energy conversion. J. Power Sources. **260**, 186–196 (2014)

Contents

Chapter 1
Introduction

Chemical energy storage is generally the method of choice for low- to medium-power applications due its high specific energy [1]. Electrochemical energy conversion and storage devices such as batteries, fuel cells, and supercapacitors are commonly utilized to access chemical energy and convert it into electrical energy at high efficiency. Electrochemical energy conversion is galvanic when electrical energy is generated by the cell reaction, while the reverse process of using electrical energy to produce chemical energy is electrolytic [2]. Fuel cells are considered galvanic electrochemical cells which convert chemical energy of a continuously supplied fuel and oxidant combination directly into electrical energy [3]. Similarly, redox flow batteries (RFBs) are electrochemical cells which utilize two redox couples dissolved in separate liquid electrolytes as the fuel and oxidant [4], featuring both galvanic and electrolytic operations. As thermodynamically open systems, fuel cells and RFBs have decoupled energy storage and conversion subsystems and can be instantly recharged with new reactants, a desirable property which eludes conventional closed-cell batteries and may offer a potential solution to the growing demand for compact energy supplies for portable electronics and a host of other low- to medium-power applications [1].

The device level power sources currently used in portable and wireless electronics are dominated by solid-state lithium ion batteries. Although the high cell voltage and energy density enabled by lithium derivatives are favorable compared to most other battery chemistries, the relatively high cost and limited resources of lithium pose certain constraints on the growth of this market. Principally, the accelerating power demands of integrated electronics are already compromised by the size, weight, and reliability of existing battery technologies. Furthermore, the manufacturing cost of mass-produced electronic devices is often dominated by the cost of the power source. Small, integrated electrochemical flow cells may potentially offer lower cost and higher overall energy density than solid-state battery systems, and the rapidly growing energy requirements are in favor of new, conceptually redesigned power packages with extended runtime. In contrast to solid-state batteries, electrochemical flow cells have fundamentally decoupled energy storage and energy

E. Kjeang, *Microfluidic Fuel Cells and Batteries*, SpringerBriefs in Energy, DOI 10.1007/978-3-319-06346-1_1, © The Author(s) - SpringerBriefs 2014

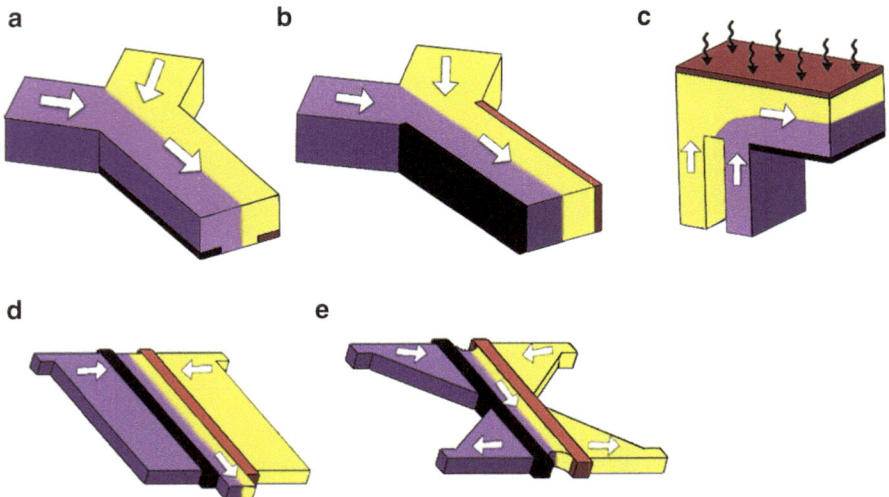

Fig. 1.1 Key microfluidic fuel cell and battery architectures developed to date: (**a**) monolithic flow-by cell; (**b**) multilayer flow-by cell; (**c**) air-breathing flow-by cell; (**d**) monolithic flow-through cell; and (**e**) monolithic dual-pass flow-through cell

conversion functions, which enables independent optimization of both subsystems as a major stride towards increased performance and functionality. There are numerous technical challenges, however, related to practical miniaturization of electrochemical flow cell technologies. For instance, hydrogen-powered polymer electrolyte fuel cells require hydrogen storage or fuel reformer units that are generally too bulky for integrated electronics. Direct liquid fuel cells and RFBs have compact fuel storage solutions, but the performance is restricted by relatively slow electrochemical kinetics and reactant crossover through the membrane that separates the two half-cells. Current membrane electrode assembly (MEA) designs are also inadequate for monolithic integration into miniaturized devices due to the laminated structure [1].

Microfluidic fuel cells, also known as membraneless fuel cells or laminar flow-based fuel cells, represent an emerging fuel cell technology capable of integration and operation within the framework of a microfluidic chip. In microfluidic fuel cells, all functions and components related to reactant delivery, reaction sites, and electrode structures are confined to a single microfluidic channel. Microfluidic fuel cells predominantly operate using co-laminar flow of fuel and oxidant electrolytes without a physical barrier, such as a membrane, to separate the two half-cells.

The microfluidic fuel cell concept was first invented and demonstrated in 2002 [5, 6]. As shown schematically in Fig. 1.1a, b, the original cell design introduced two reactants (fuel and oxidant) through separate inlet ports into a single microfluidic channel with electrodes (anode and cathode) patterned on the channel walls. In microfluidic channels, mixing of two parallel laminar streams is minimal, and reactant crossover can be avoided by strategic electrode positioning sufficiently far away from the liquid–liquid interface at the center of the channel. Following the initial breakthrough of this concept, a swell of promising microfluidic fuel cell and redox flow battery architectures have been developed. The field of microfluidic fuel cells has matured as a subset of microstructured fuel cells, or micro fuel cells, as indicated by three review articles [7–9] and a book chapter [10] on this topic. More general review articles on micro fuel cell technology are also available in the literature [11, 12].

The most influential microfluidic fuel cell architectures presented to date are illustrated schematically in Fig. 1.1. The original and most fundamental cell designs feature two streams combined horizontally in a T- or Y-channel with electrodes on the bottom (Fig. 1.1a) or side walls (Fig. 1.1b). Alternatively with the F-channel design two streams may be combined vertically in a sheath-flow configuration with electrodes situated on top and bottom walls (Fig. 1.1c). The F-channel architecture is compatible with integration of a gaseous half-cell by exposing the top electrode to the gas phase. In the flow-through porous electrode architecture (Fig. 1.1d), reactants flow through and react within porous electrodes prior to combining in a horizontal co-laminar flow in the center channel. Flow-through porous electrodes may also be applied symmetrically for reactant recirculation or recharging in microfluidic RFBs (Fig. 1.1e) with two inlets and two outlets.

Microfluidic fuel cells and batteries utilize the laminar flow characteristic of microchannels to delay convective mixing of two stratified streams carrying the fuel and oxidant, respectively. The anolyte (fuel) and catholyte (oxidant) streams contain supporting electrolyte that facilitates ionic charge transport by electromigration, thereby closing the electrical circuit of the cell. Mixing of the fuel and oxidant species is however limited to relatively slow diffusion restricted to an interfacial width at the center of the channel. In general, the electrodes are positioned on one or more walls of the manifold with sufficient spatial separation from the co-laminar interface to avoid reactant crossover. The resulting crossover rate and mixing width can be uniquely controlled by tuning of channel dimensions and flow rate.

Microfluidic electrochemical cells can mitigate some of the issues encountered in conventional MEA-based fuel cells and RFBs. For instance, the problems and costs associated with reactant gas humidification, membrane degradation, and reactant crossover are either limited or eliminated with microfluidic flow cells. It is also possible to optimize the composition of the anolyte and catholyte streams independently, thereby providing an opportunity to enhance reaction rates and cell voltage compared to incumbent membrane-based cell designs. Miniaturization of electrochemical flow cells provides the full benefits of compactness accompanied by an increase in surface-to-volume ratio, which scales as the inverse of the characteristic length. The overall performance of the surface-based electrochemical reactions therefore benefits directly from miniaturization of the device. However, the most significant benefit of microfluidic cells is the potential cost reduction. Microfluidic fuel cells and batteries are compatible with inexpensive and scalable micromachining and microfabrication methods, and the direct cost of the membrane, which is substantial for most MEA-based devices, is eliminated. Although catalyst may still be required, a variety of catalyst-free cells have been developed with carbon-based electrodes that are orders of magnitude less expensive than the platinum-containing electrodes of traditional fuel cells. Additionally, microfluidic cells do not require auxiliary systems for humidification, water management, or cooling and can be effectively operated at room temperature. The main research challenges for microfluidic fuel cells and batteries to become practical contenders, however, are in the areas of energy density and fuel utilization. While significant advances have been made, as summarized herein, these challenges remain to a large degree and will require further research.

The overall goal of this book is to provide a general overview of microfluidic fuel cell and battery technology in support of future research, product development, and commercialization activities. The focus is on emerging microfluidic fuel cell and battery devices that utilize membraneless electrochemical flow cell architectures to produce electrical power. The scope is limited to membraneless cells, as the elimination of the membrane provides compatibility with planar microfabrication and micromachining methods that are well suited for production of low-cost miniaturized power sources. However, membraneless cells require innovative microfluidic engineering solutions, which will be addressed in this work.

The present book is outlined as follows. Chapter 2 describes the fundamentals required for design and operation of microfluidic cells, followed by Chapter 3 that presents how to build and characterize the cells. Chapter 4 provides a tour of the main technological breakthroughs and contributions in this field, from the first invention of the laminar flow fuel cell to advanced high-performance cell architectures. Next, the fundamentals and devices are combined in Chapter 5, focusing on modeling advances used to inform cell design. This provides a transition into the research trends and directions (Chapter 6), intended to shed light on ongoing and emerging research activities in this field. Finally, the book is wrapped up with conclusions and recommendations (Chapter 7), including an outlook for future work towards practical commercial targets. It is hoped that the present book will entice students, researchers, and engineers alike to generate new knowledge, creativity, and innovation towards the development of practical microfluidic power sources for real-world applications.

References

1. J. Morse, Int. J. Energ. Res. **31**, 576–602 (2007)
2. A.J. Bard, L.R. Faulkner, *Electrochemical Methods: Fundamentals and Applications*, 2nd edn. (Wiley, New York, 2001)
3. M.M. Mench, *Fuel Cell Engines* (Wiley, Hoboken, NJ, 2008)
4. Z. Yang, J. Zhang, M.C.W. Kintner-Meyer, X. Lu, D. Choi, J.P. Lemmon, J. Liu, Chem. Rev. **111**, 3577–3613 (2011)
5. L.J. Markoski, E.R. Choban, J. Stoltzfus, J.S. Moore, P.A. Kenis, in *Power Sources Proceedings,* vol. 40, Cherry Hill, NJ, 2002, pp. 317–320
6. R. Ferrigno, A.D. Stroock, T.D. Clark, M. Mayer, G.M. Whitesides, J. Am. Chem. Soc. **124**, 12930–12931 (2002)
7. E. Kjeang, N. Djilali, D. Sinton, J. Power. Sources **186**, 353–369 (2009)

8. S.A. Mousavi Shaegh, N.-T. Nguyen, S.H. Chan, Int. J. Hydrogen Energ. **36**, 5675–5694 (2011)
9. B. Ho, E. Kjeang, Cent. Eur. J. Eng. **1**, 123–131 (2011)
10. E. Kjeang, N. Djilali, D. Sinton, Advances in microfluidic fuel cells, in *Micro Fuel Cells: Principles and Applications*, ed. by T.S. Zhao (Elsevier B.V., Amsterdam, 2009). ISBN 978-0-12-374713-6
11. C.K. Dyer, J. Power. Sources **106**, 31–34 (2002)
12. A. Kundu, J.H. Jang, J.H. Gil, C.R. Jung, H.R. Lee, S.H. Kim, B. Ku, Y.S. Oh, J. Power. Sources **170**, 67–78 (2007)

Chapter 2
Theory

2.1 Electrochemical Principles

Electrochemical cells are based on two half-cells, each having an electrode in contact with an electrolyte, joined electronically by external wires and ionically in the electrolyte phase in order to close the circuit. Fuel cells [1] and batteries [2] are two common types of electrochemical cells that are designed to convert chemical energy into electrical energy through electrochemical reactions at their two electrodes. In a fuel cell, which is a thermodynamically open process, a flow of externally stored fuel and oxidant is continuously supplied to the electrochemical cell to generate electrical power, in contrast to batteries where reactants and products are stored internally in a closed system without mass flux across its boundaries. A redox flow battery (RFB) [3] is an interesting device in this context, as it can be considered either a battery or a fuel cell depending on where the system boundaries are drawn. RFBs have independent energy conversion and energy storage subsystems similar to fuel cells but are generally categorized as batteries due to storage of charge within a closed liquid electrolyte system.

Both fuel cells and batteries principally comprise an anode and a cathode separated by an electrolyte, as shown schematically in Fig. 2.1 in the case of a fuel cell. The power generation function of electrochemical cells is conceptually straightforward. The fuel (or anodic reactant) is oxidized at the anode, releasing reaction products including ions and electrons. The ions travel through the electrolyte phase, which can be either a liquid or a polymer that promotes ionic conduction while insulating for electronic transport, and recombine with the oxidant (or cathodic reactant) at the cathode. The electrons required for the reduction reaction at the cathode are conducted from the anode through external wiring, thereby generating an electrical current used to drive a load. Fuel cell electrodes require contact between three separate phases at the active sites to facilitate heterogeneous electrochemical reactions that produce a useful current: the solid phase that conducts electrons to or from the electrode; the liquid or gaseous fuel or oxidant phase; and the liquid or solid polymeric electrolyte phase. In the case of a RFB, each reactant is in the liquid

E. Kjeang, *Microfluidic Fuel Cells and Batteries*, SpringerBriefs in Energy, DOI 10.1007/978-3-319-06346-1_2, © The Author(s) - SpringerBriefs 2014

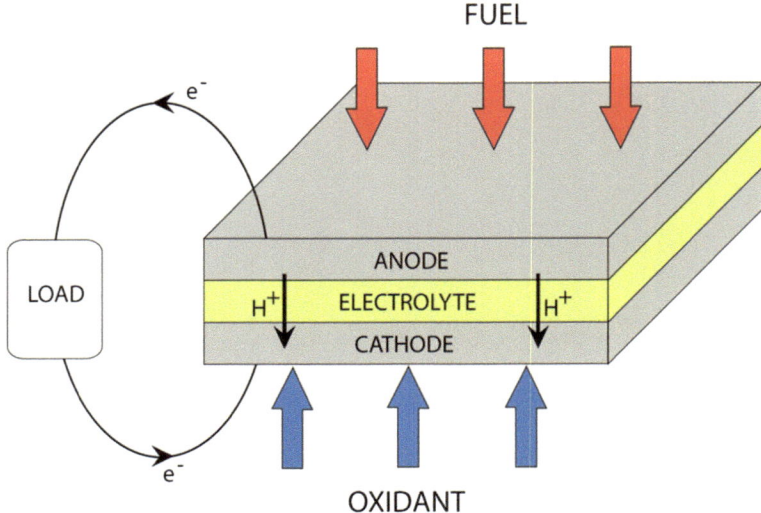

Fig. 2.1 Conceptual fuel cell layout showing the core components of the membrane electrode assembly (MEA): anode and cathode separated by a polymeric ion-conducting electrolyte and connected to an external load

phase in three-phase contact with a solid electrode and an ion-conducting liquid or polymeric electrolyte; however, the liquid reactant phase may contain supporting electrolyte to assist with ionic transport and thereby reduce the electrochemical interface to a pure solid–liquid interface. Although the electrochemical reactions are typically exothermic and therefore release energy, the reaction rates are often constrained by large activation energy that needs to be supplied for the reactions to proceed. There are three prevalent strategies to accelerate electrochemical reaction rates: (1) adding catalysts; (2) elevating the operating temperature; and (3) increasing the effective electrode area by incorporating micro- or nanostructured materials. The first two can be applied to any chemical reactions, while the third strategy is especially important in electrochemical cells due to the surface-based reactions that benefit from a high surface-to-volume ratio. Carbon-supported platinum (Pt/C) is widely utilized as electrocatalyst in low-temperature fuel cells, in particular for hydrogen and oxygen, due to its high activity. For other fuels such as formic acid and methanol, palladium, ruthenium, or various platinum alloys provide good catalytic properties. In contrast, most redox flow batteries benefit from rapid electrochemical kinetics on common carbon or graphite electrodes without any specific catalyst requirements.

The performance of electrochemical cells is normally measured in terms of cell voltage (ΔE_{cell}) and current (I). The cell voltage represents the difference in electrochemical potential between the two half-cells (cathode and anode), with a maximum at the reversible open-circuit voltage ($E_{cathode} - E_{anode}$). The reversible potential of each electrode is determined from the Gibbs' free energy of the reactants and products at their standard states via the Nernst equation:

$$E = E^0 - \frac{RT}{F} \ln \frac{\prod\limits_{products,i} a_i^{\upsilon_i}}{\prod\limits_{reactants,j} a_j^{\upsilon_j}}, \tag{2.1}$$

where E^0 is the reversible potential at standard state, R is the universal gas constant, T is the temperature, F is Faraday's constant, and a is the activity of each species ($a = 1$ at standard state), which for gaseous and aqueous species can be approximated by the partial pressure and concentration, respectively. The actual cell voltage obtained during operation is significantly lower than the reversible cell potential due to various losses (also referred to as overpotentials). The operational cell voltage is determined by [1]:

$$\Delta E_{cell} = \left(E_{cathode} - E_{anode} \right) - \left| \eta_{anode} \right| - \left| \eta_{cathode} \right| - iR_{cell} - \eta_{trans} \tag{2.2}$$

where the subtracted terms correspond to voltage losses caused by activation overpotentials due to irreversibilities at the electrodes (η), ohmic resistance of the cell (R_{cell}), and concentration overpotentials from mass transport limitations (η_{trans}). The current density (i) is the cell current divided by the geometrical surface area of the electrode. Power density (mW cm^{-2}), which is an overall measure of the device level performance, is obtained by multiplying cell voltage and current density. The electrochemical reactions and electrode materials employed in microfluidic cells are generally consistent with those of conventional electrochemical cells, and the large body of literature available on electrochemistry can be adopted for detailed descriptions of applicable reaction mechanisms, kinetics, and overpotentials [4].

2.2   Fluid Dynamics

Microfluidics is the principal subject of fluid flow on the microscale and has been described as both a science and a technology [5, 6]. It is formally defined as the study and application of fluid flow and transport phenomena in microstructures with at least one characteristic dimension in the range of 1–1,000 μm [5, 6]. The subject of microfluidics regularly involves engineering, chemistry, and biology disciplines and serves a wide range of applications including lab-on-chip technologies, biomedical diagnostics, drug discovery, proteomics, and energy conversion. Squires and Quake [7] and Gad-El-Hak [8] provide comprehensive reviews of the physics of microfluidics. Fluid flow in microscale conduits is laminar under most conditions. Flow in this regime is characterized by low Reynolds' numbers $Re = \rho U D_h / \mu$, where ρ is the fluid density, U is the average velocity, D_h is the hydraulic diameter, and μ is the dynamic viscosity. Microfluidic laminar flow is dominated by viscous effects over inertial effects, and surface forces play a dominant role over body forces.

Microfluidic electrochemical cells exploit the properties of laminar flow in microchannels to delay convective mixing of two stratified streams carrying the respective anodic and cathodic reactants. At low Re, the two streams will flow in parallel down a single microfluidic channel, as shown schematically in Fig. 1.1. This type of flow is referred to as co-laminar flow and is functional in the laminar internal flow regime for Re up to approximately 1,000. Internal flows exceeding this threshold will transition to turbulence and destabilize the co-laminar flow interface, leading to excessive mixing and loss of cell voltage.

Due to the laminar nature of microfluidics, the velocity field $\bar{u}$ for incompressible Newtonian fluids is governed by the Navier–Stokes equations for momentum conservation in 3-D:

$$\rho\left(\frac{\partial \bar{u}}{\partial t} + \bar{u} \cdot \nabla \bar{u}\right) = -\nabla p + \mu \nabla^2 \bar{u} + \bar{f}, \tag{2.3}$$

where p represents pressure and $\bar{f}$ summarizes the body forces per unit volume. The use of the Navier–Stokes equations assumes that the fluid may be treated as a continuum; however, this assumption is generally valid in microscale liquid flows [6–8] and may be applied with reasonable accuracy into the nanofluidic range. At very low Re, the nonlinear convective terms in the Navier–Stokes equations may be safely neglected, resulting in linear and predictable Stokes flow:

$$\rho\frac{\partial \bar{u}}{\partial t} = -\nabla p + \mu \nabla^2 \bar{u} + \bar{f}. \tag{2.4}$$

Furthermore, mass conservation for fluid flow obeys the continuity equation:

$$\frac{\partial \rho}{\partial t} + \nabla \cdot \left(\rho \bar{u}\right) = 0. \tag{2.5}$$

For fluids with constant density, this equation is reduced to the incompressibility condition, $\nabla \cdot \bar{u} = 0$. In classical fluid dynamics problems, e.g., flow between parallel plates and flow in a cylindrical tube, Eqs. (2.4) and (2.5) lead to the familiar parabolic pressure-driven velocity profile, which serves as a useful baseline for microfluidics.

The surface-area-to-volume ratio, which is inversely proportional to the characteristic length, is comparatively high in microfluidic devices and increases with decreasing channel dimensions. A high surface-to-volume ratio is favorable for surface-based (i.e., heterogeneous) chemical reactions such as the electrochemical reactions occurring in fuel cells and batteries. However, reducing the size of the channel leads to increasing frictional losses and parasitic load required to drive the flow. It is hence important to consider the role of pressure drop due to friction when designing microfluidic electrochemical cells. The pressure drop required to generate a pressure-driven laminar flow with mean velocity U in a straight channel of length L and hydraulic diameter D_h is conveniently expressed as [9]

$$\Delta p = \frac{32\mu L U}{D_h^2}. \tag{2.6}$$

The corresponding pumping power W required to drive the flow is obtained by multiplying the pressure drop with the flow rate Q:

$$W = \Delta p Q = \frac{32 \mu L U Q}{D_h^2}. \tag{2.7}$$

This relationship is valid for fully developed flow in a straight channel and does not include the contributions from inlet and outlet feed tubes and minor losses due to ports, bends, expansions/contractions, steps, corners, etc. In most microfluidic fuel cell designs, however, relatively long thin channels are applied where friction losses of the type described by Eq. (2.6) are known to dominate.

2.3 Transport Phenomena

Microfluidic laminar flows enable a great deal of control over fluid–fluid interfaces [10] and provide unique functionality. Most important for microfluidic electro-chemical cells is co-laminar streaming. Specifically, when two liquid streams of similar fluids in terms of viscosity and density are joined in a single microfluidic channel, a parallel co-laminar flow is established. The resulting fluid–fluid interface may be applied to observe chemical reactions in real time, serve as a lens, or sepa-rate reactants as required for microfluidic electrochemical cells. Species transport within microscale flows can occur through convection, diffusion, and electromigra-tion. In the absence of electromigration, mixing between two co-laminar streams occurs by crosswise diffusion alone. Microscale devices generally experience high Péclet numbers $Pe = UD_h/D$, where D is the diffusion coefficient. High Péclet num-bers indicate that the rate of mass transfer via crosswise diffusion is much lower than the streamwise convective velocity. In the case of microfluidic electrochemical cells, diffusive mixing is therefore restricted to a thin interfacial width at the center of the channel. This interfacial mixing width has an hourglass shape with maximum width (δ_x) at the channel walls as described by the following scaling law [11] for pressure-driven laminar flow of two aqueous solutions:

$$\delta_x \propto \left(\frac{DHz}{U} \right)^{1/3}, \tag{2.8}$$

where z is the downstream position and H is the channel height. Eq. (2.8) is limited, however, to liquids of similar density. With disparate densities, a gravity-induced reorientation of the co-laminar liquid–liquid interface can occur [12]. The physics of co-laminar flow is the key-enabling mechanism of several microfluidic devices such as the T-sensor [13], Y-mixer [14], and H-filters [15] with applications in lab-on-chip diagnostic technologies and can also be applied to selectively pattern microfluidic systems [16].

Microfluidic electrochemical cells, as shown in Fig. 1.1, employ one laminar stream that contains the fuel (or first reactant) and a second laminar stream that contains the oxidant (or second reactant). As the fuel and oxidant streams flow in a co-laminar format, the liquid–liquid interface serves as a virtual separator without the need for a membrane. The positions of the electrodes on the channel walls are however constrained by the width of the co-laminar interdiffusion zone. To prevent mixed potentials due to fuel and oxidant crossover, the electrodes must have sufficient separation from the liquid–liquid interface throughout the channel. The position and orientation of the electrodes also influence fuel utilization and overall performance of the cell. Notably however, the degree of mixing in microfluidic electrochemical cells can be effectively controlled by tuning the flow rate in the channel, vis-à-vis Eq. (2.8). Specifically, the residence time t_{res} in the electrochemical chamber must be shorter than the diffusion time t_{diff} for crossover of a reactant species to the opposite electrode in order to avoid a mixed electrode potential. This necessary criterion can be estimated using Einstein's relation for one-dimensional Brownian diffusion [17], which provides a lower bound on the average diffusion time:

$$t_{res} = \frac{L}{U} < \frac{W^2}{2D} = \overline{t_{diff}}.$$ (2.9)

This equation includes two channel size parameters, namely the channel length L and width W (the mean diffusion distance). The mean velocity is determined by dividing the flow rate Q by the channel height H and width W. These additional constraints can be included to obtain a useful dimensionless relation for the ratio of solute (reactant) advection to cross-stream diffusion in co-laminar flow cells:

$$\frac{QW}{2DHL} > 1.$$ (2.10)

A similar design rule can also be derived for this geometry from the Péclet number [18], which is the ratio between advective flux and diffusive flux. From this point of view, Eq. (2.10) is satisfied when the rate of downstream advective transport exceeds the rate of reactant crossover towards the opposite electrode. As the relation is derived for an average diffusion time, it is recommended for the ratio to exceed 1 by an appreciable margin. For instance, to accommodate a suggested minimum ratio of 10 for a species with a $\sim 10^{-10}$ m^2 s^{-1} diffusion coefficient in a 0.5 mm high, 0.5 mm wide, and 10 mm long channel, the required flow rate is on the order of ~ 1 μL min^{-1}.

Both co-laminar streams must have relatively high ionic conductivity to facilitate good ionic charge transport between the electrodes and to close the electrical circuit. High conductivity is normally provided by the addition of a supporting electrolyte that contains ions with high mobility, e.g., hydronium or hydroxide ions. The supporting electrolyte also stabilizes the co-laminar flow with respect to electromigration of fuel and oxidant species, since it is these highly mobile constituents that redistribute and shield the effects of the electric field and electric double layers in the channel. The ohmic resistance for ionic transport in the channel can be expressed in terms of the average charge-transfer distance between the electrodes (d_{ct}), the cross-sectional area for charge transfer (A_{ct}), and the ionic conductivity (σ) as follows:

$$R_f = \frac{d_{ct}}{\sigma A_{ct}}.$$ (2.11)

Equation (2.11) indicates that a strong supporting electrolyte with high ionic conductivity and a high aspect ratio rectangular microchannel with closely spaced electrodes are therefore desired. This design strategy is partially in conflict with that required for efficient separation of fuel and oxidant. Principally, the interdiffusion width according to Eq. (2.8) indicates a lower limit on the electrode spacing. Striking an adequate balance between the competing requirements for species transport and ionic conductivity is essential in microfluidic electrochemical cells. Ohmic resistance in these cells is generally higher than in MEA-based fuel cells due to the additional constraints for cell design. Increasing the concentration of the supporting electrolyte is a convenient mitigation approach to achieve higher conductivity than for ionomer membranes. Ultimately, however, the choice of supporting electrolyte should be made with consideration of optimum reaction kinetics. The co-laminar configuration uniquely permits the composition of the two streams to be chosen independently, thus providing an opportunity to improve reaction rates and cell voltage. Similarly, the cell potential can be increased through adjusting the reversible half-cell potentials by pH modification of the individual streams.

Reactant transport from the bulk flow to the electrode surface takes place primarily by convection and diffusion in the absence of significant electromigration, provided a strong supporting electrolyte is used. In this case, species conservation takes the general form:

$$\nabla \cdot \left(C_i \bar{u} \right) = -\nabla \cdot \bar{J}_i + R_i, \tag{2.12}$$

where C_i is the local concentration of species i and R_i is a source term that describes the net rate of generation or consumption of species i via homogeneous chemical reactions. Under the infinite dilution assumption, the diffusive flux of species i is calculated by Fick's law:

$$\bar{J}_i = -D_i \nabla C_i \tag{2.13}$$

where D_i is the diffusion coefficient of the species i in the appropriate medium. Heterogeneous electrochemical reactions at the electrode surfaces are the boundary conditions of Eq. (2.13). The ratio of reaction rate and mass transport rate is defined by the Damköhler number (Da). When a current is drawn, a concentration boundary layer develops over the electrode starting at the leading edge. Assuming the electrochemical reactions are rapid (high Da), the maximum current density of a microfluidic electrochemical cell is determined by the rate of the convective/diffusive mass transport from the bulk to the surface of the electrode. In this transport limited case, the reactant concentration is zero at the entire surface of the electrode. Kjeang et al. [19] provided scaling laws for microfluidic fuel cell operation in the transport-controlled regime based on pseudo-3D flow over a flat plate, using previously developed theory for electrochemical flow sensors [20] originating from the classical Graetz problem of heat transfer [21]. The analysis is based on dimensionless formulations of mean velocity and current:

$$U^* = \frac{U D_h^2}{lD}, \tag{2.14}$$

$$I^* = \frac{I}{nFdlc_0 D / D_{\mathrm{h}}}.$$

$$(2.15)$$

There are two distinct regimes: (1) the high U^* regime, which covers most practical flow rates experienced in microfluidic fuel cells and (2) the low U^* regime for low flow rates in channels with small hydraulic diameter. In the high U^* regime, the transport limited current is proportional to the cubic root of the mean velocity [20]:

$$I^* = -1.849 U^{*1/3}.$$

$$(2.16)$$

In the low U^* regime, the flux of reactant entering the channel is equal to the rate of the electrochemical reactions, such that all reactant molecules are converted into product species and useful current. In this case, the maximum current is directly proportional to inlet concentration and flow rate [20]:

$$I = -nFc_0 Q.$$

$$(2.17)$$

Again, we can use the dimensionless quantities I^* and U^* to derive a relationship that is valid for high aspect ratio channels under any conditions within the low U^* regime:

$$I^* = -\frac{1}{2} U^*.$$

$$(2.18)$$

The transition point between the high U^* regime and the low U^* regime occurs at $U^{*1/3} = 1.9$ [20]. The coulombic single-pass fuel utilization can also be defined in this context as the rate of reactant consumption by the electrochemical reactions divided by the flux of reactant supplied by the flow:

$$\varepsilon_f = \frac{I}{nFc_0 Q}.$$

$$(2.19)$$

More generally, there is no single dominating limiting factor, and the current density of a microfluidic cell is controlled by a combination of mass transport, electrochemical kinetics, and ohmic resistance. This trio of potential limiting factors must be considered when designing a new device. The overall single-pass energy conversion efficiency of microfluidic electrochemical cells is defined by the product of the coulombic efficiency (fuel utilization) and voltage efficiency. In the case of galvanic cells, the energy conversion efficiency for discharging is written as

$$\varepsilon = \varepsilon_f \cdot \varepsilon_v = \frac{I}{nFc_0 Q} \cdot \frac{E_{\mathrm{cell}}}{E_{\mathrm{cell}}^\circ},$$

$$(2.20)$$

where E_{cell}° represents the theoretical, reversible cell potential. This potential is sometimes replaced by the thermodynamic cell potential, E_{th}, to calculate the thermodynamic efficiency of a fuel cell. In the case of electrolytic cells, the voltage efficiency term is reversed, and the energy conversion efficiency for charging is given by

$$\varepsilon = \varepsilon_f \cdot \varepsilon_v = \frac{I}{nFc_0 Q} \cdot \frac{E_{\mathrm{cell}}^\circ}{E_{\mathrm{cell}}}.$$

$$(2.21)$$

This equation assumes that all applied current contributes to the desired electrolytic cell reaction; if parasitic side reactions are present that consume a portion of the applied current, such additional coulombic losses must also be accounted for in the coulombic efficiency.

In addition, the parasitic pumping power requirements to drive the flow (Eq. (2.7)) must be kept substantially below the power produced by the cell. In principle, any power consumption required for cell operation ought to be accounted for in the overall system efficiency of the device. Microchannels with ~μL to ~mL per minute flow rates generally provide an optimum balance with respect to the above-mentioned constraints.

References

1. J. Larminie, A. Dicks, *Fuel Cell Systems Explained* (Wiley, Hoboken, 2003)
2. R.M. Dell, D.A.J. Rand, *Understanding Batteries* (Royal Society of Chemistry, London, 2001)
3. D.P. de Leon, A. Frias-Ferrer, J. Gonzalez-Garcia, D.A. Szanto, F.C. Walsh, J. Power. Sources **160**, 716–732 (2006)
4. W.M. Haynes (ed.), *CRC Handbook of Chemistry and Physics*, 94th edn (Taylor & Francis/ CRC Press, New York, 2013)
5. G.M. Whitesides, Nature **442**, 368–373 (2006)
6. N.T. Nguyen, S.T. Wereley, *Fundamentals and Applications of Microfluidics* (Artech House, Boston, MA., 2002)
7. T.M. Squires, S.R. Quake, Rev. Mod. Phys. **77**, 977–1026 (2005)
8. M. Gad-el-Hak, Phys. Fluids **17**, 100612 (2005)
9. C.T. Crowe, D.F. Elger, J.A. Roberson, *Engineering Fluid Mechanics* (Wiley, New York, 2001)
10. J. Atencia, D.J. Beebe, Nature **437**, 648–655 (2005)
11. R.F. Ismagilov, A.D. Stroock, P.J.A. Kenis, G. Whitesides, H.A. Stone, Appl. Phys. Lett. **76**, 2376–2378 (2000)
12. S.K. Yoon, M. Mitchell, E.R. Choban, P.J.A. Kenis, Lab Chip **5**, 1259–1263 (2005)
13. A.E. Kamholz, B.H. Weigl, B.A. Finlayson, P. Yager, Anal. Chem. **71**, 5340–5347 (1999)
14. J.B. Salmon, C. Dubrocq, P. Tabeling, S. Charier, D. Alcor, L. Jullien, F. Ferrage, Anal. Chem. **77**, 3417–3424 (2005)
15. J.P. Brody, P. Yager, Sensor. Actuator. Phys. **58**, 13–18 (1997)
16. P.J.A. Kenis, R.F. Ismagilov, G.M. Whitesides, Science **285**, 83–85 (1999)
17. A. Einstein, *Investigations on the Theory of the Brownian Movement*, vol. 17, 2nd edn. (Dover Publications, Mineola, NY, 1956)
18. R. Byron Bird, W.E. Stewart, E.N. Lightfoot, *Transport Phenomena*, 2nd edn. (Wiley, New York, 2001)
19. E. Kjeang, J. McKechnie, D. Sinton, N. Djilali, J. Power. Sources **168**, 379–390 (2007)
20. E. Kjeang, B. Roesch, J. McKechnie, D.A. Harrington, N. Djilali, D. Sinton, Microfluid. Nanofluid. **3**, 403–416 (2007)
21. L. Graetz, Über die Wärmeleitungsfähigkeit von Flüssigkeiten. Ann. Physik. **25**, 337–357 (1885)

Chapter 3
Fabrication and Testing

3.1 Fabrication

Fabrication methods developed originally for electronics, and later microfluidics, have been applied to good effect in the area of microfluidic electrochemical cells. Most commonly, these devices consisted of a microchannel, a pair of electrodes, and a liquid-tight support structure as shown schematically in Fig. 3.1. Microchannels have generally been fabricated by rapid prototyping, using standard photolithography and soft lithography protocols [1, 2], also known as micromolding. The channel structures were commonly molded in poly(dimethylsiloxane) (PDMS) and subsequently sealed to a solid substrate. PDMS has relatively benign properties for electrochemical applications; it is relatively inert and compatible with most solvents and electrolytes [3]. Electrodes were either prepatterned on the substrate or inserted into prefabricated grooves in the channel layer. Alignment of the channel structure with the electrode pattern can be facilitated by a suitably modified mask aligner, if available, or by hand. Indeed, for lab-on-a-chip devices which typically consist of a single microfluidic layer sealed to a substrate, soft lithography is a prevailing approach due to its rapid prototyping capabilities and low cost. In the case of microfluidic fuel cells and batteries, soft lithography enables monolithic, membraneless planar device architectures that are well suited for integration into on-chip systems and other MEMS devices.

A convenient and inexpensive soft lithography-based procedure for fabrication of microfluidic electrochemical cells is summarized as follows. The procedure consists of three processes: the creation of a master, the molding of a channel structure, and the assembly and bonding of the microfluidic cell. The purpose of the master is to define a positive microchannel pattern for subsequent molding into a soft polymer to create the microchannel part of the cell. A cleaned and pretreated substrate such as a microscope glass slide or a silicon wafer is coated with a thin layer of photoresist by spin coating. The coated substrate is then baked on a hot plate to stabilize the photoresist. The substrate is then exposed to collimated UV light through a photomask window that defines the desired channel structure

E. Kjeang, *Microfluidic Fuel Cells and Batteries*, SpringerBriefs in Energy,
DOI 10.1007/978-3-319-06346-1_3, © The Author(s) - SpringerBriefs 2014

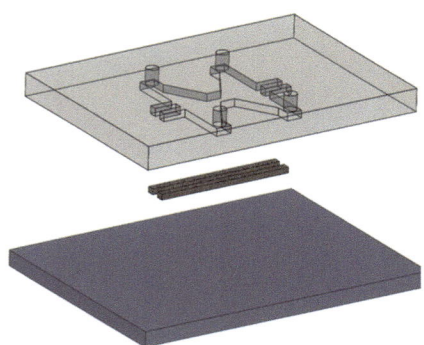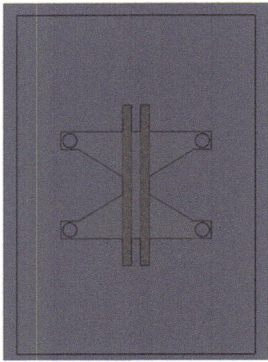

Fig. 3.1 Conceptual schematic of the main parts (microchannel layer, electrodes, and substrate) and assembly (*top view*) of a microfluidic electrochemical cell

Fig. 3.2 Photograph of a master featuring a positive ridge pattern (150 μm high) created in photoresist on a silicon wafer prepared for replica molding into a soft polymer

(typically drafted in CAD software and printed on a transparency by a high-resolution image setter). After an additional bake that polymerizes the exposed portion of the photoresist, the unexposed portion is removed by immersion of the substrate in developer liquid. The result is a master with a positive pattern defined by the remaining polymerized photoresist ridges, as illustrated in Fig. 3.2. This master may be reused many times, depending on the quality of the original coating and the intricacy and aspect ratio of the features.

Next, the channel structure is molded by pouring a liquid polymer (often PDMS) over the master, followed by degassing in vacuum and subsequent curing on a hot plate. The polymer part containing a negative imprint of the channel structure is then cut from the mold and removed from the master. The obtained channel structure is then sealed to a substrate, typically glass or PDMS, either reversibly as is or irreversibly following plasma treating of both parts. Plasma treating can be conveniently performed using a corona discharge coil, as shown in Fig. 3.3. As an additional benefit, this process renders the PDMS channel walls hydrophilic, which

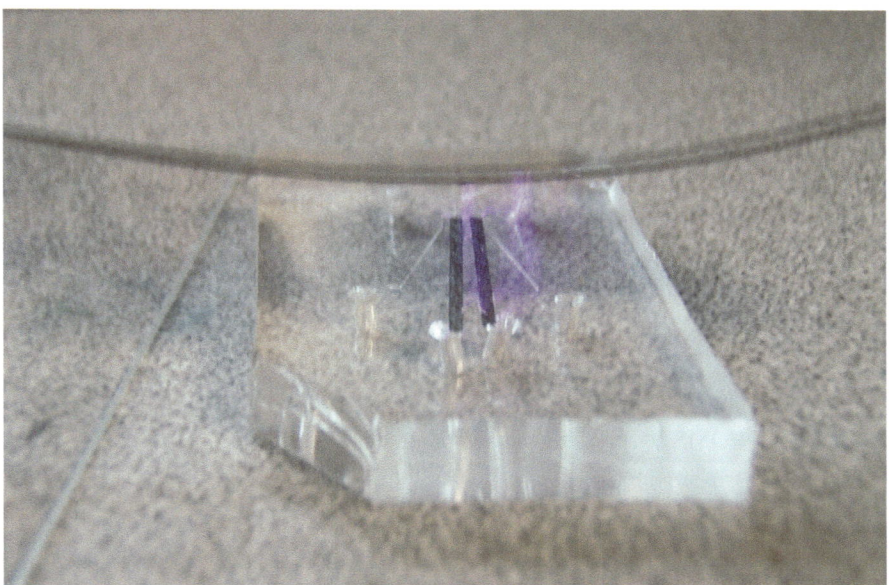

Fig. 3.3 Plasma treating of a PDMS channel structure (with two electrodes inserted) using a corona discharge coil

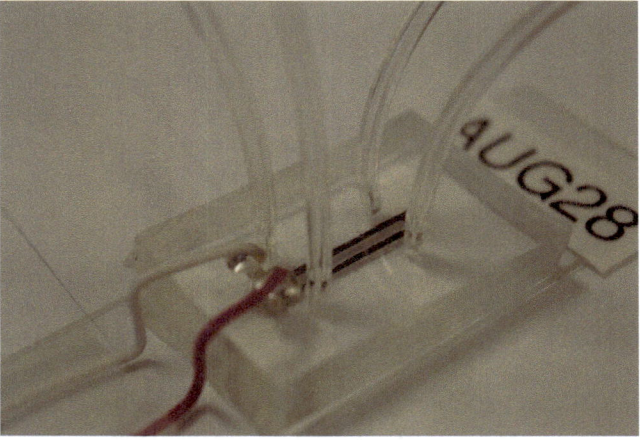

Fig. 3.4 Sample photograph of an assembled microfluidic electrochemical cell ready for testing and experimentation

promotes wetting and reduces pressure drop in the channel. Access holes are punched in the PDMS to create fluid ports and electrical contacts, where required. Upon assembly of the plasma-treated channel structure and substrate parts, a permanent bond is created that provides a sufficiently strong seal for liquid-tight cell operation without physical clamping of the device. Finally, the assembled microfluidic cell is interfaced with tubes and electrical wires, as depicted in Fig. 3.4.

Microchannel structures for microfluidic cells can also be fabricated by photo-lithographic techniques directly in the photoresist material [4, 5]. In this case, the channel is made in negative relief. Alternatively, four separate parts, each contributing one channel wall, may be assembled to form the microfluidic channel. This method was employed by Choban et al. [6, 7] to incorporate electrodes on the side walls of a channel with horizontal streaming (Fig. 1.1b). Two graphite plates were aligned using separators (spacers) and vertically sealed with PDMS films and polycarbonate capping layers. The part count can potentially be reduced to three using a channel stencil with an open area defining the channel and its two side walls. The stencil approach is practical for microfluidic cells employing vertically layered streaming and top and bottom walls provided by electrode substrates. The channel height is determined by the thickness of the stencil. For instance, PDMS channel stencils of ~1 mm height have suitable mechanical strength for handling and can be fabricated in PDMS by pressing a solid plate against the ridge pattern. Jayashree et al. [8, 9] fabricated air-breathing microfluidic fuel cells by joining a 1 mm high poly(methylmethacrylate) (PMMA) stencil to a graphite plate anode and a porous gas diffusion cathode. Alternatively, laser etching of PMMA enables precise control of the geometry by adjusting the speed and power of the laser beam and is generally very rapid (~ few seconds). The PMMA stencil of Li et al. [10] was bound between two top and bottom Au electrode-patterned PMMA parts using adhesive gaskets, also prepared by laser cutting.

Microfluidic fuel cells have also been constructed using micromachining methods [11, 12]. Originating from the integrated circuit chip industry, silicon has become the dominating substrate material for micromachining technologies. A silicon substrate is patterned by a series of lithography steps and wet- or dry-etched to form the desired structures. Finally, the substrate is bonded to glass or additional silicon wafers for air tight sealing. A silicon-based microfluidic fuel cell was developed by Cohen et al. [12]. In brief, a silicon channel stencil was created using potassium hydroxide wet-etching of a photoresist-patterned silicon wafer. The obtained silicon stencil was coated with an insulator material to prevent electrical short-circuiting and subsequently sealed between two flexible polyamide electrode films. The most important advantage of micromachining in the context of microfluidic electrochemical cells is its potential for significant cost reduction contributed by batch processes for high-volume manufacturing [13].

Microfluidic fuel cells have also been built using standard, widely available machine shop equipment. CNC machining [14] was used to construct a millimeter scale cavity in a block of Delrin plastic for application in three-dimensional microfluidic fuel cell arrays. The CNC technique is convenient and fast although the resolution and general applicability is limited to relatively large features and components. Notably, the recent emergence of 3-D printers can have game-changing impact on this field. As of 2013, however, the commonly available 3-D printing technology has insufficient dot resolution and chemical compatibility for reliable construction of microfluidic cells. Nevertheless, 3-D printing of microfluidic fuel cells was recently demonstrated and successfully applied to dimensional scale-up of single-cell devices [15].

Electrode positioning and patterning is an equally important consideration for reliable design and construction of microfluidic electrochemical cells. Most microfluidic cells developed to date employed patterned electrodes positioned in parallel on the bottom wall of the channel. The bottom substrate, e.g., glass, was commonly patterned by lift-off or etching-based photolithography. The lift-off approach uses a photoresist layer on the substrate with a negative imprint of the desired electrode pattern. The photoresist-patterned substrate is then coated with a conductive material such as graphite, gold, and platinum over an adhesive layer (e.g., chromium or titanium) by standard evaporation or sputtering techniques. The photoresist layer is then removed, hence the name lift-off, leaving only the desired electrode pattern. The etch approach utilizes a similar strategy, where the substrate is first uniformly coated with a conductive layer followed by a positive photoresist layer pattern. An etching step removes the conductive material, except for the electrode pattern protected by the photoresist. The remainder of the photoresist is then removed to reveal the desired pattern. Commercial prefabricated gold slides can be used as an alternative to the initial evaporation or sputtering processes [16]. Otherwise, rigid, self-contained electrode structures such as graphite rods [14] and carbon paper strips [17–19] have also been employed as both electrodes and current collectors in cells with horizontal streaming using embossing or custom-grooved PDMS. Graphite rods have been additionally employed as structural elements as well as current collectors and electrodes [8]. The electrocatalytic properties and surface area of the electrodes can be improved by application of a first or additional layer of conductive material and/or catalyst. Commonly utilized techniques include electrodeposition [4, 5, 16, 20–22], ink coating or spraying [6–8], electron-beam evaporation [5, 11, 12, 21, 22], sputtering [23, 24], and micromolding [25]. The surface roughness and chemical composition of electrode surfaces are critical to the performance of electrochemical devices and microfluidic fuel cells and batteries are no exception in this regard.

Sealing is a common concern in microfluidic devices. The most common way to accomplish a liquid-tight seal during operation of a multilayer microfluidic cell is to physically clamp the assembled parts together using PMMA or aluminum end plates. This approach works well when gaskets are used or with internal parts made of smooth elastomeric materials, e.g., PDMS. The advantage of a mechanically enforced seal is that it may be easily undone following testing to salvage different components. The use of a clamping device, however, requires additional parts and space. Alternatively, an irreversible seal may be achieved between PDMS/PDMS and PDMS/glass by plasma treating of both parts prior to assembly, as previously described.

3.2 Testing

The primary goal of testing a microfluidic fuel cell or battery is to obtain a current/voltage (*I/V*) curve, also known as polarization curve, which is the most commonly used performance metric in this field. Testing and characterization of microfluidic electrochemical cells requires fluidic and electronic interfaces facilitated on-chip by ports and contacts, respectively. Externally, these interfaces are connected via tubes and wires to various hardware tools and instrumentation systems employed to control and operate the devices. In the most common configuration, a syringe pump is applied to drive the reactant flow while a potentiostat is utilized as an electronic load. Using a syringe pump is convenient for many reasons, including precise control of flow rate, wide range of flow rates, and support of different syringe sizes. Moreover, most modern analytical syringe pumps feature an electronic display, touch screen, and/or computer interface and are designed to hold multiple syringes which can support multi-stream microfluidic cell operation. Alternatively, the fluidics can be controlled by means of gravity by raising the reactant reservoirs to higher elevation than that of the outlet waste fluid compartment. However, accurate flow rate control is more challenging with the gravity method, as any change in pressure drop due to elevation change or bubble formation during an experiment will directly influence the flow rate.

The electronic part of an electrochemical flow cell experiment is normally operated using a potentiostat. Potentiostats are equipped with a wide range of electronic instrumentation and software capabilities to suit any experimental need in the general area of electrochemistry. The range of current supported by potentiostats is compatible with testing of small microfluidic fuel cells and batteries with low total power output. For testing of large single cells and stacks of multiple cells, however, larger electronic loads are generally recommended. In a standard performance test, the working and sensing leads are connected to the anode while the counter and reference leads are connected to the cathode such that both cell voltage and current can be simultaneously controlled and measured. Both galvanostatic and potentiostatic experiments are regularly performed in this configuration and it is good practice to ensure that both techniques lead to the same results before systematic characterization is conducted with either method. During each measurement, it is important to wait for both flow rate and current/voltage to reach steady state, which can take several minutes. Linear scan voltammetry or amperometry can be conveniently applied to measure polarization curves using low scan rates (e.g., 1 mV s^{-1}). Most modern potentiostats or frequency response analyzers (FRAs) can also be employed to measure impedance spectra of microfluidic electrochemical cells by applying a small sinusoidal voltage perturbation and measure the resulting AC current signal across a range of frequencies. This method is often used to measure the combined ohmic cell resistance of microfluidic electrochemical cells by extracting the high-frequency real-axis intercept of the Nyquist plot of impedance [26]. Current densities and power densities (obtained by multiplying current density and cell voltage) are calculated by normalizing the current by the geometric (projected) area of the electrodes, which is the general norm established for conventional electrochemical cells.

The in situ performance of individual electrodes (anode and cathode) within microfluidic cells can be characterized independently by using a separate reference electrode. However, with electrodes in the 100–400 µm thickness range, the height of the microfluidic channel is generally too small to insert a liquid junction reference electrode. Therefore, reference electrodes are usually placed in the outlet fluid reservoir, and positioned as close as possible to the exit of the microchannel. Saturated calomel or Ag/AgCl electrodes are most often used because the relatively isolated liquid junction preserves the potential of the electrodes throughout the experiment, making them useful as a stable reference point from which to make measurements. This method has successfully been utilized to measure single electrode polarization curves in situ for a variety of anode and cathode half-cells inside the microfluidic fuel cell domain [7, 18]. Notably, a functional reference electrode at the outlet requires a continuous electrolyte pathway to the working electrode and is therefore unsuitable for half-cells that involve gas phase reactants or products. Impedance analysis of individual half-cells can also be carried out using this method [26].

For educational purposes, the potentiostat can effectively be replaced by a standard multimeter and a student-designed load bank consisting of a breadboard, a set of resistors, and wires—all of which are usually available in a standard teaching laboratory for undergraduate level electronics courses. The polarization curve can be obtained point-by-point by measuring the voltage across each resistor and calculating the corresponding current using Ohm's law.

References

1. J.C. McDonald, D.C. Duffy, J.R. Anderson, D.T. Chiu, H.K. Wu, O.J.A. Schueller, G.M. Whitesides, Electrophoresis 21, 27–40 (2000)
2. D.C. Duffy, J.C. McDonald, O.J.A. Schueller, G.M. Whitesides, Anal. Chem. 70, 4974–4984 (1998)
3. J.N. Lee, C. Park, G.M. Whitesides, Anal. Chem. 75, 6544–6554 (2003)
4. R. Ferrigno, A.D. Stroock, T.D. Clark, M. Mayer, G.M. Whitesides, J. Am. Chem. Soc. 124, 12930–12931 (2002)
5. W. Sung, J.-W. Choi, J. Power. Sources 172, 198–208 (2007)
6. E.R. Choban, J.S. Spendelow, L. Gancs, A. Wieckowski, P.J.A. Kenis, Electrochim. Acta 50, 5390–5398 (2005)
7. E.R. Choban, P. Waszczuk, P.J.A. Kenis, Electrochem. Solid State Lett. 8, A348–A352 (2005)
8. R.S. Jayashree, D. Egas, J.S. Spendelow, D. Natarajan, L.J. Markoski, P.J.A. Kenis, Electrochem. Solid State Lett. 9, A252–A256 (2006)
9. R.S. Jayashree, L. Gancs, E.R. Choban, A. Primak, D. Natarajan, L.J. Markoski, P.J.A. Kenis, J. Am. Chem. Soc. 127, 16758–16759 (2005)
10. A. Li, S.H. Chan, N.T. Nguyen, J. Micromech. Microeng. 17, 1107–1113 (2007)
11. J.L. Cohen, D.J. Volpe, D.A. Westly, A. Pechenik, H.D. Abruna, Langmuir 21, 3544–3550 (2005)
12. J.L. Cohen, D.A. Westly, A. Pechenik, H.D. Abruna, J. Power. Sources 139, 96–105 (2005)
13. S.D. Senturia, Microsystem Design (Kluwer, Dordrecht, 2001)
14. E. Kjeang, J. McKechnie, D. Sinton, N. Djilali, J. Power. Sources 168, 379–390 (2007)
15. B. Ho, Master's thesis, Simon Fraser University, Scale up Solutions for Liquid Based Microfluidic Fuel Cell, 2012
16. E. Kjeang, A.G. Brolo, D.A. Harrington, N. Djilali, D. Sinton, J. Electrochem. Soc. 154, B1220–B1226 (2007)

17. E. Kjeang, R. Michel, D. Sinton, N. Djilali, D.A. Harrington, Electrochim. Acta **54**, 698–705 (2008)
18. E. Kjeang, R. Michel, D.A. Harrington, N. Djilali, D. Sinton, J. Am. Chem. Soc. **130**, 4000–4006 (2008)
19. E. Kjeang, B.T. Proctor, A.G. Brolo, D.A. Harrington, N. Djilali, D. Sinton, Electrochim. Acta **52**, 4942–4946 (2007)
20. E.R. Choban, L.J. Markoski, A. Wieckowski, P.J.A. Kenis, J. Power. Sources **128**, 54–60 (2004)
21. S.M. Mitrovski, L.C.C. Elliott, R.G. Nuzzo, Langmuir **20**, 6974–6976 (2004)
22. S.M. Mitrovski, R.G. Nuzzo, Lab Chip **6**, 353–361 (2006)
23. M. Togo, A. Takamura, T. Asai, H. Kaji, M. Nishizawa, Electrochim. Acta **52**, 4669–4674 (2007)
24. S. Hasegawa, K. Shimotani, K. Kishi, H. Watanabe, Electrochem. Solid State Lett. **8**, A119–A121 (2005)
25. C.M. Moore, S.D. Minteer, R.S. Martin, Lab Chip **5**, 218–225 (2005)
26. F.R. Brushett, R.S. Jayashree, W.-P. Zhou, P.J.A. Kenis, Electrochim. Acta **54**, 7099–7105 (2009)

Chapter 4
Devices

As of 2013, the research advances in the field of microfluidic fuel cells and batteries have included more than 100 scientific publications, and the technology is currently being developed for commercial and military applications (on a limited scale) by several private–public R&D partnerships including INI Power Systems (Morrisville, NC) with intellectual property licensed from the University of Illinois at Urbana-Champaign, Laminare Technologies (Ithaca, NY) in collaboration with Cornell University, and IBM (Zurich, Switzerland) in collaboration with Ecole Polytechnique Fédérale de Lausanne (EPFL). Most microfluidic electrochemical cell developments leverage reactant chemistries adopted from conventional membrane or membrane electrode assembly-based fuel cells and redox flow batteries. Prototype microfluidic fuel cell and battery devices have been demonstrated based on a wide range of fuels, including hydrogen, methanol, formic acid, glucose, glycerol, sodium borohydride, hydrazine, ethanol, hydrogen peroxide, and vanadium redox species. Several microfluidic fuel cell concepts have also been demonstrated for biofuel cells, including microfluidic bioanodes based on ethanol and glucose fuel. Oxygen in aqueous or gaseous form is the most frequently used oxidant, followed by hydrogen peroxide, vanadium and cerium redox species, potassium permanganate, bromine, and sodium hypochlorite in aqueous media. Most devices incorporated a supporting electrolyte within the reactant streams to promote ionic charge transport between the electrodes and reduce ohmic resistance. This feature was accommodated by adding a strong acid or base, e.g., sulfuric acid or potassium hydroxide, with highly mobile and soluble ionic components. The co-laminar flow in the microchannel was generally implemented using horizontal streaming with vertical liquid–liquid interface or vertical sheath flow with horizontal liquid–liquid interface. In the case of horizontal streaming, the cells had T-, Y-, Ψ-, or H-shaped microchannel designs with electrodes positioned in parallel on the bottom wall (Fig. 1.1a), on opposite side walls (Fig. 1.1b), or in cross-flow configuration (Fig. 1.1d, e). For cells using vertical sheath flow, the channels were predominantly F-shaped with electrodes positioned on the top and bottom walls (Fig. 1.1c). In addition, certain

E. Kjeang, *Microfluidic Fuel Cells and Batteries*, SpringerBriefs in Energy, DOI 10.1007/978-3-319-06346-1_4, © The Author(s) - SpringerBriefs 2014

specialized microfluidic fuel cells with selective catalysts and stable reactants do not require co-laminar flow to delay mixing and related crossover effects. In this special case, the reactant species can be mixed in a single I-shaped stream [1–3].

The invention of the laminar flow-based fuel cell in 2002 was followed by two early microfluidic fuel cell demonstrations that provided the foundation for future technology advances in this field [4, 5]. The pioneering cell introduced by Choban et al. [5] featured a horizontal Y-shaped microchannel with Pt-coated electrodes on the side walls, housing an aqueous HCOOH fuel stream and an aqueous O_2 saturated oxidant stream. The power output of this device was restricted by the rate of mass transport to the active sites, primarily in the cathodic half-cell, and the overall system performance suffered from low fuel utilization. The cathodic transport limitation was confirmed by switching to potassium permanganate oxidant with higher solubility in aqueous media, resulting in an order of magnitude higher power density [5]. The high-conductivity liquid electrolyte employed enabled the use of an external reference electrode to characterize individual half-cells and measure ohmic resistance in situ during fuel cell operation. With this experimental approach, the overall cathodic mass transport limitation of dissolved O_2-based cells was verified [6]. For early devices using formic acid in the anodic stream and dissolved oxygen in the cathodic stream, the measured power density of ~0.2 mW cm^{-2} [5, 7] was primarily constrained by the low solubility (1–4 mM) and diffusivity (2×10^{-5} cm^2 s^{-1}) of O_2 in the aqueous electrolyte and by CO poisoning of the Pt catalyst used for HCOOH oxidation. Bimetallic Pt/Ru nanoparticles for methanol oxidation that are less susceptible to CO poisoning were able to substantially raise the power density to ~3 mW cm^{-2} [6]. An additional benefit of nanoparticle catalysts is the high electrocatalytic surface area (roughness factor of ~500) that further promotes the electrochemical kinetics. Alternatively, adatoms of Bi can be adsorbed on Pt to reduce CO poisoning and improve cell performance, as shown by Cohen et al. [7]. More recently, the general microfluidic fuel cell design has been systematically employed by Arriaga and coworkers to demonstrate new advances in fuel chemistry and catalyst development. For instance, liquid phase glucose [8] and glycerol [9] were introduced as practical alternative fuels, while advanced catalyst supports featuring multiwalled carbon nanotubes were applied to enhance the performance of the more commonly used formic acid anode [10, 11].

4.1   Co-laminar Mixed Media Streaming

The co-laminar flow principles of microfluidic electrochemical cells enable mixed media operation, in contrast to traditional types of fuel cells and redox flow batteries operating under all-acidic or all-alkaline conditions imposed by the membranes. The unique mixed media capability allows independent tuning of half-cell conditions for optimization of reaction kinetics and cell potential. In mixed media conditions, the open-circuit cell voltage can be increased by shifting the reversible half-cell potentials via pH modification of individual streams. For instance, the reversible potential (E) of the oxygen reduction reaction (ORR)

$$O_2 + 4H^+ + 4e^- \leftrightarrow 2H_2O \tag{4.1}$$

depends on pH according to the Nernst equation:

$$E = E^0 - \frac{RT}{F} \ln \frac{\prod_{products,i} a_i^{v_i}}{\prod_{reactants,j} a_j^{v_j}} = E^0 - \frac{RT}{F} \ln \frac{1}{a_{O_2}\left(a_{H^+}\right)^4}, \tag{4.2}$$

where E^0 is the reversible potential at standard state and a the activity of each species ($a = 1$ at standard state), which for aqueous species can be approximated by the molar concentration. By reducing the pH of the catholyte (adding more H^+), the reversible potential of the cathode becomes more positive and thereby yields larger open-circuit cell voltage. Similarly, the anode potential can be made more negative by using alkaline electrolytes.

Cohen et al. [12] demonstrated that the open-circuit potential of a microfluidic H_2/O_2 fuel cell can be raised well beyond the standard cell potential of 1.23 V using mixed media. An alkaline dissolved hydrogen stream and an acidic dissolved oxygen stream were implemented in a co-laminar microfluidic fuel cell of the previously reported F-shaped architecture [7]. An increased cell potential was obtained from the negative shift of the hydrogen oxidation potential in the alkaline environment. The power produced in the dual electrolyte configuration was more than doubled compared to the corresponding single electrolyte systems despite the competing effect of relatively slow hydrogen oxidation kinetics in alkaline media. The media flexibility was also studied by Choban et al. [13] by operating a microfluidic methanol/oxygen fuel cell under all-acidic, all-alkaline, and mixed media conditions. The membraneless cell design eliminated the issue of membrane clogging by anodic carbonate products formed in alkaline media, which is otherwise common with many conventional direct methanol fuel cells. Interestingly, it was determined that alkaline conditions had positive effects on the reaction kinetics at both electrodes. In addition, the cell potential was increased by several hundred millivolts under mixed media conditions with alkaline anolyte and acidic catholyte up to an impressive 1.4 V. A peak power density of 5 mW cm^{-2} at 1.0 V was achieved with the methanol/oxygen fuel cell under mixed media conditions, compared to 2.4 and 2.0 mW cm^{-2} for all-acidic and all-alkaline conditions, respectively. Furthermore, it was observed that at cell voltages below 0.8 V, the ORR at the highly acidic cathode was complemented by proton reduction to hydrogen, thereby providing a useful mitigating effect to the cathodic mass transport-controlled cell current. This outcome was a direct consequence of the mixed media configuration, and a significantly enhanced fuel cell performance with a peak power density of 12 mW cm^{-2} was achieved. Notably, Hasegawa et al. [14] also used the mixed media approach to operate a microfluidic fuel cell in which hydrogen peroxide served as primary reactant in both anode and cathode half-cells, employing alkaline and acidic media, respectively. The direct hydrogen peroxide fuel cell produced relatively high power densities up to 23 mW cm^{-2}. The cell performance was principally constrained by the spontaneous hydrogen peroxide decomposition on the cathode and associated oxygen gas evolution that may perturb the co-laminar flow interface and lead to excessive mixing of the two streams. Moreover, the mixed media capability of microfluidic fuel cells has been strategically exploited to evaluate a wide range of fuel and oxidant chemistries in alkaline/acidic electrolyte combinations that were otherwise not possible with standard MEA-based fuel cell devices [15].

One possible caveat of microfluidic fuel cell operation under mixed media conditions is the potential for depletion or consumption of supporting electrolyte. For instance, mixed media operation may cause exothermic neutralization of OH$^-$ and H$^+$ at the co-laminar flow interface which generates a liquid junction potential and locally reduced ionic strength that can reduce cell potential and increase ohmic cell resistance. Additionally, the overall cell reaction may include net consumption of supporting electrolyte. In the alkaline/acidic mixed media cases described above, hydroxide ions and protons were consumed at the anode and cathode, respectively, meaning that the primary ionic charge carriers were also utilized as secondary reactants and were at least partially depleted in the cell. The role of the supporting electrolyte must therefore be carefully considered in the design of mixed media microfluidic cells.

4.2 Microfluidic Fuel Cells with Biocatalysts

Biocatalysts are a promising alternative to traditional catalysts, particularly in the context of microfluidic fuel cells. Fuel cells that utilize biological entities such as enzymes and microbes to catalyze the chemical reactions, thereby replacing traditional electrocatalysts, are collectively termed biofuel cells [16, 17]. The name biofuel cells is somewhat of a misnomer as these cells are not restricted to biofuels; the usage of this name is, however, prevalent. In conventional biofuel cells, biocatalytic entities are placed in a two-compartment electrochemical cell containing buffer solution with fuel and oxidant in the anolyte and catholyte compartments, respectively. In most configurations, these compartments are separated by an ion-exchange membrane or a salt bridge [16] and include redox couples acting as diffusional electron mediators (or cofactors), which is necessary for efficient catalyst utilization. The rate of electron transfer is generally determined by the rate of diffusion of these cofactors and the ion permeability of the membrane that separates the two compartments [17]. Only certain types of enzymes with active sites located on the periphery of the enzyme are capable of direct electron transfer to the electrode without redox mediators [18]. Modern biofuel cell technologies benefit from the functionalization of electrode surfaces and immobilization of active enzymes in order to improve electron transfer characteristics and stability. These approaches include covalent polymer tethering of cofactor units to multilayered enzyme array assemblies, crosslinking of affinity complexes formed between redox enzymes and immobilized cofactors on functionalized conductive supports, and noncovalent coupling by hydrophobic/ hydrophilic or affinity interactions [19] or direct encapsulation of enzymes in hydrogel films. Upon conformational fixation of the enzyme attributed to its immobilization, the stability (lifetime) may be extended from a few days in solution to weeks or even months. Interestingly, recent developments have also included advances in reversible redox chemistry towards rechargeable enzymatic biobatteries [20].

Biofuel cells are highly compatible with microfluidic electrochemical cell architectures, and most advantages associated with microfluidic fuel cells are transferable to biofuel cell technologies. Biofuel cells with nonselective electrochemistry, i.e., cells using diffusional redox mediators, can utilize the established co-laminar microfluidic fuel cell design, which enables the tailoring of independent anolyte and catholyte compositions for optimum enzymatic activity and stability. Electrodes based on immobilized enzymes and localized cofactors may also be employed in microfluidic fuel cell designs. Full selectivity of both anodic and cathodic half-cells with co-immobilized redox relays allows microfluidic biofuel cell operation in a single microchannel without the need for co-laminar flow. In this configuration, initially mixed fuel and oxidant would flow together in a single channel with species-specific oxidation and reduction occurring at the respective biocatalyst electrodes.

Despite the many advantages and opportunities offered by microfluidic cell designs, relatively few microfluidic biofuel cell works have been presented to date [21]. The area was pioneered by Moore et al. [3] through the introduction of a microchip-based bioanode with NAD-dependent alcohol dehydrogenase enzymes immobilized in a tetrabutylammonium bromide-treated Nafion membrane. The bioanode was assembled on a micromolded carbon electrode prepatterned on a glass substrate and sealed under a standard PDMS microchannel used to deliver the fuel solution containing ethanol and NAD^+ in phosphate buffer. When operated versus an external Pt cathode, the microfluidic bioanode generated an open-circuit voltage of 0.34 V and a maximum current density of 53 $\mu A\ cm^{-2}$, expected to be limited by the rate of diffusion of NADH within the membrane. Subsequently, an integrated microfluidic biofuel cell technology was developed based on a similar enzyme immobilization technique [22, 23]. The technology is currently licensed to Akermin Inc. (St Louis, MO) under the trademark "stabilized enzyme biofuel cells."

A microfluidic bioanode based on vitamin K_3-mediated glucose oxidation by the glucose dehydrogenase enzyme was developed by Togo et al. [2]. The bioanode was immobilized inside a fluidic chip containing a PDMS-coated conventional Pt cathode with an integrated Ag/AgCl reference electrode used for in situ electrochemical characterization. The bioanode was positioned downstream of the Pt cathode to minimize contamination. The flow cell produced 32 $\mu W\ cm^{-2}$ at 0.29 V when running on air-saturated pH 7-buffered fuel solution containing glucose and NAD^+. The current density of the proof-of-concept cell declined by 50 % over 18 h of continuous operation due to swelling effects. The bioanode flow cell formed the conceptual basis for the complete microfluidic biofuel cell depicted in Fig. 4.1 [24]. In this case, the Pt cathode was replaced with a bilirubin oxidase-adsorbed biocathode, and

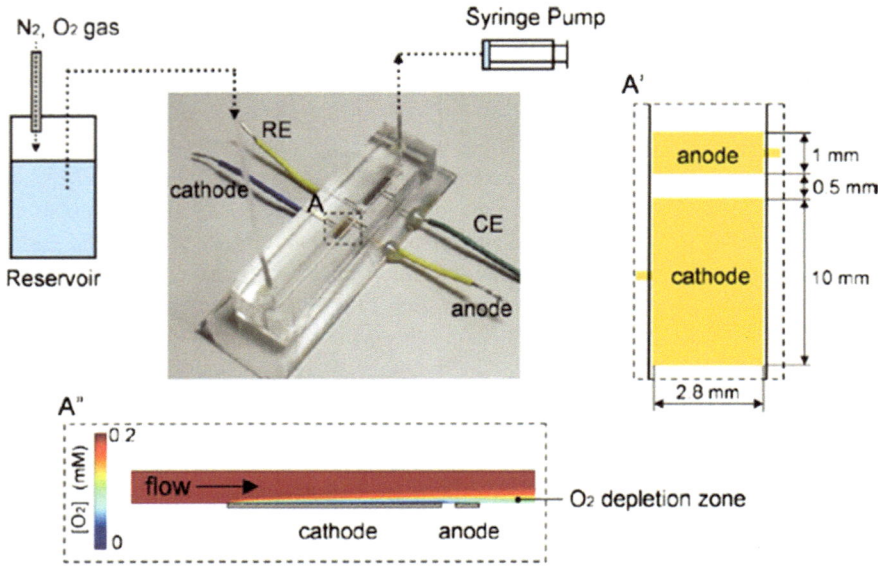

Fig. 4.1 I-shaped microfluidic biofuel cell with mixed reactant supply of an oxygen-saturated glucose solution. The highlighted reaction zone (**a**) is magnified to illustrate the electrode configuration (**a'**), with the biocathode located upstream from the bioanode, and the growth of the oxygen concentration boundary layer formed on the cathode (**a"**). Reproduced with permission from Togo et al. [24]. Copyright Elsevier (2008)

the power output of the biofuel cell was comparable to the previous device. A parametric study of flow rate, channel height, and electrode geometry demonstrated restricted access of dissolved oxygen to the biocathode, similarly to the non-biological microfluidic fuel cells previously described. The present cell design mitigated this limitation by enlarging the cathode area to ten times the anode size.

A membraneless microfluidic biofuel cell using the fungal enzyme laccase as biocatalyst for oxygen reduction and 2,2′-azinobis (3-ethylbenzohiazoline-6-sulfonate) (ABTS) as a redox mediator was developed by Lim et al. [25]. This device successfully demonstrated the use of microfluidic co-laminar flow of anolyte and catholyte supplied to an integrated biofuel cell. Parametric studies were conducted to determine the impact of electrode length and spacing on the cell performance. It was found that splitting a single electrode into two or more smaller electrodes and separating them by a sufficient distance can increase the maximum power density by 25 % compared to a single electrode configuration with identical electroactive area. A similar enzymatic biofuel cell using co-laminar flow of glucose and dissolved oxygen in a Y-shaped microfluidic channel was reported by Zebda et al. [26]. At the anode, the glucose was oxidized by glucose oxidase (GOD) with $Fe(CN)_6^{3-}$, whereas at the cathode, the oxygen was reduced by laccase in the presence of ABTS as a reduction mediator. The anolyte consisted of GOD and $Fe(CN)_6^{3-}$ in neutral phosphate buffer at pH 7, while the catholyte solution was laccase and ABTS in citrate buffer at pH 3. The assembled biofuel cell produced a maximum power density of 110 $\mu W\ cm^{-2}$ at 0.3 V and demonstrated the feasibility of independently tuned co-laminar mixed media streaming to optimize enzyme activity in biofuel cells. The performance achieved with this device is the highest reported to date in the field of microfluidic biofuel cells. Moreover, the use of pyrolyzed photoresist film electrodes as a substrate for immobilization of enzymes in microfluidic biofuel cells has shown promising performance in the case of bioelectrode-based enzymatic cells and demonstrated compatibility with silicon foundry manufacturing technologies amenable to mass production [27].

A significant research opportunity in the context of microfluidic biofuel cells is in the application of microbial fuel cells that utilize microbes or other microorganisms to extract electrons from a fuel, in this case referred to as substrate. When compared to enzymatic biofuel cells, microbial biofuel cells have the advantage of enhanced stability and lifetime due to the more robust configuration of living cells protected by the cell membrane, provided the environmental conditions are kept within appropriate levels for microbial cultures. A proof-of-concept co-laminar flow-based microbial microfluidic biofuel cell was recently demonstrated by Ye et al. [28]. The device employed a microbial biofilm coated on the anode of the co-laminar Y-channel in a membraneless multilayer architecture comprising two graphite electrodes compressed between PMMA plates. The power densities obtained were comparable to those of enzymatic microfluidic biofuel cells and thereby demonstrated the viability of the technology.

4.3   Gas Diffusion Electrodes

The oxygen solubility and transport limitations, common to many microfluidic fuel cells discussed thus far, may be addressed by incorporating cathodes that are exposed to the surrounding air. Ambient air has four orders of magnitude higher diffusivity (0.2 $cm^2\ s^{-1}$) and several times higher concentration (10 mM) than dissolved oxygen in aqueous media [29]. The concept of air-breathing electrode (GDE) was adapted from conventional electrochemical cells such as polymer electrolyte fuel cells [30] and metal-air batteries [31] and innovatively applied to the membraneless microfluidic electrochemical cells. By replacing the solid cathode electrode with a porous hydrophobic GDE that allows gaseous reactants to access the active sites while confining the liquid electrolyte to the internal microchannel, the reduction reaction can utilize oxygen from the surrounding ambient air or a separate air or oxygen gas supply, as depicted in Fig. 1.1c. Although the inlets could potentially be positioned to form a T- or Y-junction [32, 33], the F-shaped microchannel junction with the inlets on the same side has most frequently been used [15, 29, 34–37], which results in vertical sheath flow. To facilitate ionic transport to the cathodic reaction sites and sufficient separation between the co-laminar mixing interface and the cathode, the air-breathing cell architecture requires a blank cathodic electrolyte stream generally flowing in parallel with the anolyte (fuel) stream. Consequently, there is no immediate net energy density advantage for this configuration.

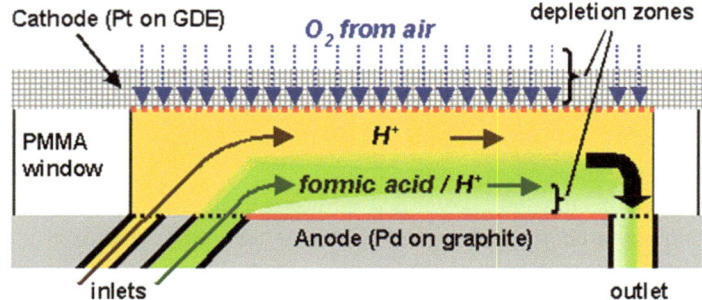

Fig. 4.2 Schematic of an air-breathing microfluidic fuel cell with F-shaped microchannel. The cathode is designed to capture oxygen from the ambient air using a gas diffusion electrode in contact with a blank electrolyte stream on the opposite side. Reprinted with permission from Jayashree et al. [29]. Copyright 2005 American Chemical Society

Jayashree et al. [29] unveiled the first microfluidic fuel cell with an integrated air-breathing cathode, featuring a graphite plate anode coated with Pd black nanoparticles and a porous carbon paper cathode coated with Pt black nanoparticles. The F-shaped fuel cell design is illustrated schematically in Fig. 4.2. With this cell a peak power density of 26 mW cm^{-2} was achieved using 1 M formic acid in 0.5 M sulfuric acid anolyte and a blank 0.5 M sulfuric acid catholyte flowing at 0.3 mL min^{-1} per stream. This performance level is comparable to the equivalent MEA-based fuel cell designs operated at room temperature and demonstrated the viability of GDEs in membraneless cells. The air-breathing cell architecture was also evaluated using methanol [35], which enables higher overall energy density than formic acid. Relatively modest power densities were obtained with 1 M methanol fuel (17 mW cm^{-2}); however, improved reaction kinetics facilitated by the membraneless design resulted in an increase in the open-circuit cell voltage from 0.93 to 1.05 V. The air-breathing cells also enabled significantly higher coulombic fuel utilization than the previous generation cells based on dissolved oxygen, up to a maximum of 33 % [29]. The performance and stability of the air-breathing cathode in a microfluidic fuel cell was enhanced by introducing multiwalled carbon nanotube catalyst support [38]. Other developments involving the air-breathing design were contributed by Shaegh et al. with the added benefits of a flow-through anode [33] and an integrated fuel reservoir [36]. In both cases, the multilayer fabrication method illustrated in Fig. 1.1c is preferred over the planar lithographic fabrication method due to the relative ease of incorporating a GDE into a sandwich structure. In either case, however, two electrolyte streams are necessary and therefore most standard channel structures (Fig. 1.1) could potentially be converted into an air-breathing cell.

Tominaka et al. [39] developed the first monolithic microfluidic fuel cell with air-breathing capabilities. The planar architecture of the silicon-based device is shown schematically in Fig. 4.3. In this case, a singular fuel and electrolyte containing I-shaped microchannel is employed that is completely open to the ambient air on the top side and in contact with two cathodes on the side walls and an anode on the

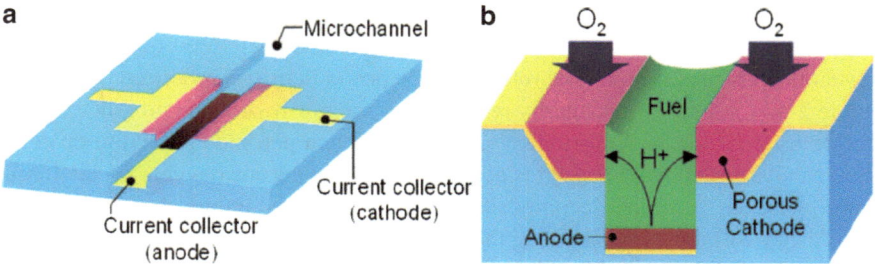

Fig. 4.3 Schematic of a monolithic, air-breathing microfluidic fuel cell with two porous ORR-selective cathodes, indicating (**a**) current collector layout and (**b**) cross-sectional 3-D view of the device. Reproduced with permission from Tominaka et al. [39]. Copyright 2008 American Chemical Society

bottom wall. This configuration provides air-breathing access to oxidant at the porous cathodes and utilized capillary forces to supply fuel to the anode and provide electrolyte contact with the cathodes. An ORR selective Pd–Co catalyst was employed on the cathodes to mitigate the direct contact with the fuel solution in the channel. The liquid fuel was contained in the microchannel by capillary forces; however, there is some potential for fuel evaporation. Modified versions of this unique microfluidic fuel cell design was shown to be bendable when fabricated on a flexible polymer substrate [40] and compatible with a variety of fuel chemistries [41].

The scale-up and integration of multiple air-breathing fuel cells, while ensuring sufficient oxidant access, can be complicated by geometrical restrictions. INI Power Systems (Morrisville, NC) is developing a direct methanol microfluidic fuel cell system with integrated gas diffusion cathode for commercial applications. Notably, recent improvements of electrodes and catalysts, optimization of methanol concentration and flow rates, and the addition of a gaseous flow field on the cathode side have resulted in impressive power densities on the order of 100 mW cm^{-2} [42]. As compared to conventional MEA-based direct methanol fuel cells, these microfluidic fuel cells are competitive.

4.4 Liquid Oxidants

Another avenue towards improved performance of mass transfer-controlled microfluidic electrochemical cells is the use of alternate reactants that are soluble in aqueous media at higher concentrations than dissolved oxygen. There are several common liquid fuels available with relatively high specific energy density, e.g., methanol, formic acid, and sodium borohydride, that are generally paired with oxygen or air cathodes when deployed for electrochemical energy conversion. With the exception of hydrogen peroxide, which was previously employed as an oxidant in direct sodium borohydride/hydrogen peroxide fuel cells [43–45], liquid oxidants are less common.

An acidic hydrogen peroxide oxidant solution was paired with formic acid fuel in a laser-micromachined microfluidic fuel cell device [46] based on an F-shaped co-laminar channel design. The relatively low power densities produced by this cell (up to 2 mW cm^{-2}) were primarily restricted by the low hydrogen peroxide concentration used (10 mM). Unfortunately, direct hydrogen peroxide reduction on common catalysts such as Pt and Pd is accompanied by vigorous oxygen gas evolution from hydrogen peroxide decomposition that must be accommodated by strategic cell design without compromising the stability of the co-laminar flow. A common problem in all-liquid microfluidic systems, the formation of bubbles within a microchannel, is often sufficient to perturb the co-laminar interface and increase reactant crossover, reduce ionic conductivity, or block the channel completely [47]. Even with liquid reactants it is possible to form gaseous products or intermediates with certain fuel and oxidant combinations. If such combinations are desired, reliable strategies must be developed to accommodate the gas phase and continuously purge it from the cell. At low current densities, dissolution of gas bubbles within the liquid electrolytes may be sufficient to remove the reaction products without compromising the co-laminar flow. At moderate to high current densities, gas bubbles can potentially be captured in purposefully grooved microchannels to minimize the disturbance [48]. Hur et al. [49] recently proposed and demonstrated that gas bubbles can be effectively applied to drive the capillary flow of electrolytes. In either case, however, gaseous products require increased system complexity and lead to decreased performance stability. Hence, reactants with gaseous intermediates or products should generally be avoided in the microfluidic electrochemical cell application due to the availability of alternate reactants.

Several approaches to stabilize the co-laminar liquid–liquid interface have recently been proposed. These stabilization schemes include magnetically separated streams [50], integration of a third electrolyte stream [51], and utilization of a grooved microchannel geometry that serves as a guide for gaseous products [48]. In the magnetic field-induced approach developed by Aogaki et al. [50], a magnetic virtual wall was deployed to precisely separate a paramagnetic oxidant solution from a diamagnetic fuel solution at a liquid–liquid interface. A zinc/copper electrochemical flow cell was successfully operated under the influence of a magnetic field provided by a permanent magnet. This approach enables rapid removal of bubbles and solid particles and precise crossover control. The previously established Y- or T-shaped co-laminar flow cell was revised into a Ψ-shape by Sun et al. [51]. The third electrolyte stream, situated in the center of the co-laminar flow, contained blank electrolyte to promote the separation of the fuel and oxidant streams and prevent interfacial reaction. The cell voltage and current density of the prototype fuel cell operated on formic acid and potassium permanganate could be optimized via precise flow rate control of the central electrolyte stream. Kjeang et al. [48] reported a microfluidic fuel cell based on formic acid and hydrogen peroxide, featuring a grooved microchannel design. Steady operation without crossover issues was demonstrated across a wide range of flow rates, and practical power densities up to 30 mW cm^{-2} were achieved. In this cell design, the microchannel was grooved over each electrode as to effectively capture all gas bubbles formed by the electrochemical reactions and prevent the otherwise destabilizing effect that bubble formation has on co-laminar flow.

The rate of gas evolution during hydrogen peroxide reduction largely depends on the choice of catalyst. Au has been identified as a promising catalyst for hydrogen peroxide reduction capable of minimizing gas evolution while still providing high current density [52]. Au catalyst is also effective for direct borohydride oxidation with limited hydrogen evolution on the anode [52]. The end product in this case, sodium metaborate, is highly soluble in aqueous media. The feasibility of coupling sodium borohydride and hydrogen peroxide in a co-laminar microfluidic fuel cell is however restricted by chemical stability issues. Specifically, sodium borohydride is only stable in alkaline media, and hydrogen peroxide requires an acidic environment to prevent fast decomposition. Co-laminar mixing of an alkaline borohydride solution with an acidic peroxide solution leads to vigorous gas formation and heat generation. Although microfluidic fuel cells that combine these two components in a single channel are unlikely, individual half-cells based on either alkaline borohydride or acidic peroxide may find useful applications in cell architectures that accommodate low rates of gas evolution, such as the previously described grooved microchannel design.

An alternative approach to mitigate the stability problem associated with gas evolution is to employ selective catalysts on both electrodes. In this case, crossover is not a concern, and the fuel and oxidant reactants can be mixed in a single I-shaped laminar stream. The mixed reactant approach is however restricted to fuel and oxidant pairs that do not react spontaneously upon mixing and have sufficiently high kinetics in a singular electrolyte. Sung and Choi [1] demonstrated a single-stream microfluidic fuel cell in alkaline solution, having a nickel hydroxide anode for methanol oxidation and a silver oxide cathode for hydrogen peroxide reduction. Due to incomplete selectivity of the present catalysts, the result was a low open-circuit voltage (0.12 V) and low power density (0.03 mW cm^{-2}).

Liquid phase reactant chemistries originally developed for redox flow batteries can be exploited to great effect in microfluidic electrochemical cells. Most commonly, vanadium redox flow battery technology utilizes soluble vanadium redox couples in both half-cells for regenerative electrochemical energy storage units [53]. The combination of aqueous redox pairs in vanadium redox cells, V^{2+}/V^{3+} and VO^{2+}/VO_2^+, provides many benefits for operation of microfluidic electrochemical cells: high reactant solubility with viable redox concentrations up to 5.4 M [54], well-balanced electrochemical half-cells in terms of both transport characteristics and reaction rates, high open-circuit voltages up to ~1.7 V at uniform pH due to the large difference in formal redox potentials, and catalyst-free electrochemical reactions facilitated by plain carbon electrodes. Accordingly, the first journal publication in the emerging field of microfluidic fuel cells was an all-vanadium microfluidic redox fuel cell introduced by Ferrigno et al. in 2002 [4]. The original proof-of-concept cell featured a Y-shaped microchannel with planar graphite-covered gold electrodes patterned on the bottom wall and generated comparatively high power densities up to 38 mW cm^{-2}. These power levels however required a high flow rate of 1.5 mL min^{-1}

per stream and the fuel utilization was limited to ~0.1 %. A conceptually similar vanadium redox fuel cell based on graphite rod electrodes was developed by Kjeang et al. [55]. Graphite rods, commonly employed in mechanical pencils, are inexpensive and provide combined electrodes and current collectors in a single unit with high electrical conductivity. The prototype graphite rod fuel cell delivered useful power density levels at high flow rates. In addition, the high aspect ratio (width/height) cross-sectional geometry of the microchannel enabled fuel cell operation at low flow rates with substantially improved fuel utilization up to 63 % per single pass. A cell voltage breakdown analysis revealed that the performance was principally constrained by convective/diffusive species transport from the bulk fluid. Hence, the cell design was further improved by replacing the graphite rods with integrated porous carbon electrodes [56], resulting in increased active area and enhanced transport characteristics. A peak power density of 70 mW cm^{-2} was achieved with the microfluidic vanadium redox fuel cell with porous electrodes.

4.5 High-Performance Microfluidic Cell Architectures

The overall performance of microfluidic electrochemical cells is generally dictated by reactant mass transport limitations. It is also the unique mass transport characteristics of microfluidic cells that distinguish them from the conventional MEA-based cell designs. Therefore, mass transport in microfluidic electrochemical cells is both an interesting and useful subject for research and has spearheaded the contributions towards enhanced cell performance over the past several years.

Microfluidic cells generally rely on cross-stream diffusion to transport reactants from the bulk co-laminar flow to the active sites of the electrodes. Diffusion on the microscale is often relatively slow and this leads to the common transport limitations discussed earlier. While the inner structure of porous electrodes provides increased surface area and aids diffusive species transport, regular microfluidic fuel cells with solid electrodes replaced by porous electrodes fail to take full advantage of these structures. Kjeang et al. [57] modified the vanadium redox fuel cell architecture by sealing the porous electrodes between the top and bottom substrates in order to force the reactants to cross directly through the electrodes. The proposed flow-through porous electrode cell is shown schematically in Fig. 1.1d, in profile in Fig. 4.4a, and in operation in Fig. 4.4b, c. As shown, the two cross-flow reactant streams meet in an orthogonally arranged central microchannel where they are directed towards the outlet in a co-laminar format. Due to the disparity in flow resistance between the channel and the porous electrodes, the flow distribution was uniform through the entire porous electrodes. This aspect enabled utilization of the full depth of the porous medium and associated active area, and provided enhanced species transport from the bulk flow to the active sites. The various colors inherent to different vanadium redox species in solution enable convenient visualization of the fuel cell operation. At open circuit (Fig. 4.4b), the central channel exhibits a co-laminar flow of unused fuel (purple) and oxidant (black). In operation at 0.8 V

Fig. 4.4 (**a**) Schematic and (**b–d**) annotated images of a microfluidic fuel cell with flow-through porous electrodes. The images show regular fuel cell operation at (**b**) open-circuit and (**c**) 0.8 V cell voltage and (**d**) regenerative operation running in reverse at 1.5 V applied voltage. The colors of the vanadium electrolytes indicate V^{2+} (*purple*) and V^{3+} (*light green*) at the anode and VO_2^+ (*black*) and VO^{2+} (*light blue*) at the cathode. Reproduced and adapted with permission from Kjeang et al. [57]. Copyright 2008 American Chemical Society

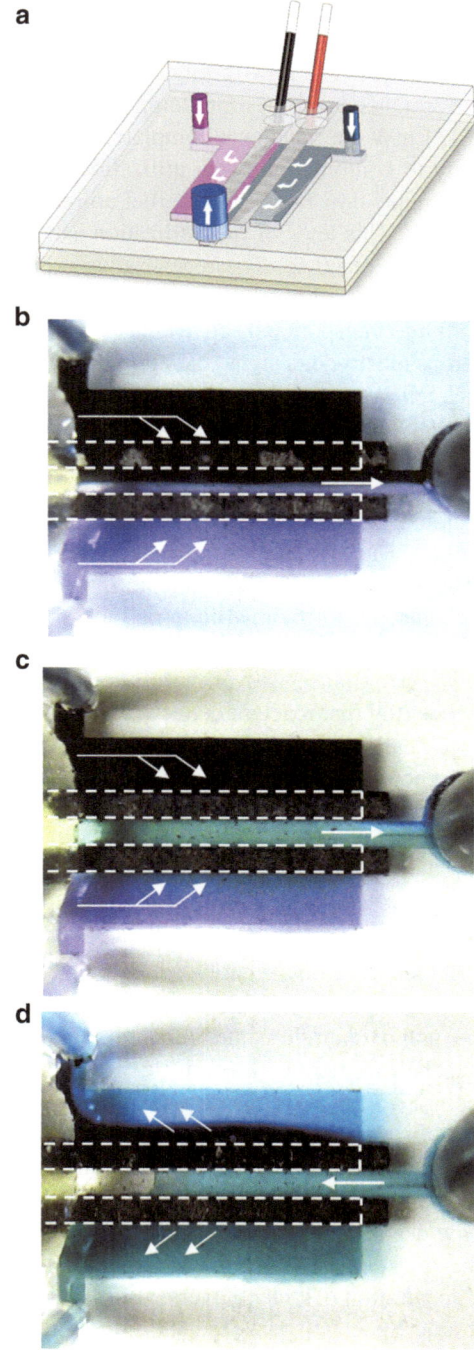

(Fig. 4.4c), the fuel and oxidant are largely consumed in the porous electrode structures prior to reaching the central channel, as indicated by the uniform, light blue and green colors. The flow-through porous electrode architecture enabled class-leading performance levels at room temperature, including steady state power densities up to 131 mW cm^{-2} and near complete fuel utilization. The fuel cell also had the capability to combine high fuel utilization with high cell voltages. As an example, at 1 μL min^{-1} flow rate an active fuel utilization of 94 % per single pass was achieved at 0.8 V. This level of fuel utilization is equivalent to an overall energy conversion efficiency of 60 %. The regenerative capabilities of this cell (Fig. 4.4d) are discussed in Sect. 4.6.

While transport and active area aspects significantly improved with flow-through porous electrode designs, the main factor limiting the overall energy density of the fuel cell system is the use of vanadium redox couples and their limited solubility. One strategy to address this issue is to modify and operate a flow-through porous electrode cell in alkaline mode using formate fuel and hypochlorite oxidant [58]. Reactant solutions of formic acid and sodium hypochlorite are both available at low cost and stable as highly concentrated liquids. Formate oxidation and hypochlorite reduction in alkaline media on porous Pd and Au electrodes were shown to have rapid kinetics at low overpotentials while preventing gaseous CO_2 formation by carbonate absorption. The prototype formate/hypochlorite fuel cell with flow-through porous electrode architecture delivered a peak power density of 52 mW cm^{-2}. This performance was primarily constrained by high ohmic cell resistance. It is noteworthy that reactant concentrations well below the solubility limits were used. Specifically, the hypochlorite solution employed was that of commercially available household bleach.

A creatively designed and promising microfluidic fuel cell architecture with flow-through porous electrodes was developed by Salloum et al. [59]. As illustrated in Fig. 4.5, this cell operates using sequential radial flow-through concentric porous electrodes. The anolyte flow enters through the center of a disc-shaped anode and flows radially towards a ring-shaped cathode. The partially consumed anolyte is then blended with a catholyte stream in a small ring-shaped cavity prior to entering the porous cathode. The concentric cell design enables independent control over the fuel and oxidant flow rate, although the impact of the fuel crossover to the cathode must be considered. The performance of concentric cells stands to benefit from high fuel utilization at the anode and/or the use of selective catalysts on the cathode.

Mass transport in microfluidic cells can also be enhanced through active boundary layer control. Specifically, the concentration boundary layers that arise due to reactant conversion in operating microfluidic fuel cells can be replenished via strategic design modifications as an approach to improve the overall fuel cell performance. Yoon et al. [60] recommended three key strategies for active control of concentration boundary layers, as illustrated schematically in Fig. 4.6: (a) removing product species (spent reactants) through multiple periodically placed outlets; (b) inserting fresh reactants through multiple periodically placed inlets; and (c) generating a secondary transverse flow induced by custom-designed topographical herringbone patterns on the channel walls. The effectiveness of these approaches was

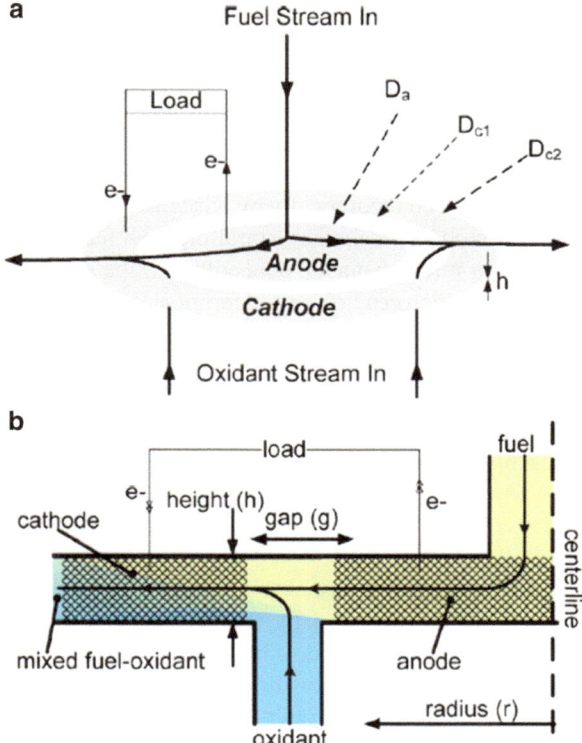

Fig. 4.5 A radial flow, concentric microfluidic fuel cell with flow-through porous electrodes shown schematically in (**a**) isometric projection and (**b**) cross-sectional view. Reproduced with permission from Salloum et al. [59]. Copyright Elsevier (2008)

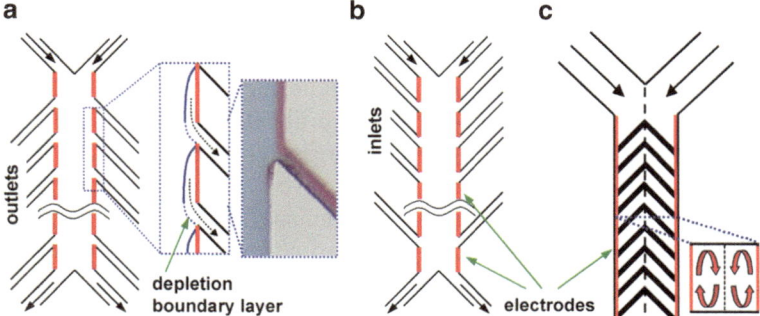

Fig. 4.6 Active control of concentration boundary layers in microfluidic cells by (**a**) removing consumed species through multiple periodically placed outlets; (**b**) adding reactants through multiple periodically placed inlets; and (**c**) generating a secondary transverse flow by topographical herringbone patterns on the channel walls. Reproduced from Yoon et al. [60] by permission of the Royal Society of Chemistry (RSC)

evaluated through numerical simulations and experimental verification using the ferri-/ferrocyanide redox couple as a model system. A chronoamperometric study indicated that by adding two extra inlets, the transport limited cell current could be enhanced by ~30 % without increasing the net flow rate.

The use of topographical herringbone ridge patterns was first proposed by Stroock et al. [61] in an effort to increase cross-stream mixing of species for lab-on-chip applications. As described theoretically by Kirtland et al. [62] for microreactor systems, these patterns induce a secondary spiralling flow that increases the rate of cross-stream transport in microchannels, depending on the exact geometry of the pattern. In the case of a microreactor, an experimental demonstration showed a 10–40 % increase in current density with the provision of herringbone ridges on one wall. For electrochemical reactions such as those occurring in microfluidic fuel cells, this approach can potentially be employed to increase both current density and fuel utilization. However, in devices relying on precisely controlled co-laminar flow of reactants, caution must be exercised to avoid a simultaneous increase in fuel and oxidant mixing at the co-laminar flow interface. Additional trade-offs include increased fabrication complexity and/or increased parasitic pumping power required to drive the flow.

Lim and Palmore [25] proposed passive boundary layer control by splitting the electrodes into smaller units separated by a gap and demonstrated the concept experimentally using a microfluidic redox fuel cell with five sets of consecutive electrodes. The measured current density at the second set of electrodes was found to increase by the passive replenishment of the concentration boundary layer in the gap section. However, since the geometrical area of the gap did not contribute any additional current, the device-level current density was not improved. A numerical optimization analysis of the same concept was conducted by Lee et al. [63], recommending arrays of miniaturized electrodes, i.e., nanoelectrodes, in a similar configuration. Generalized design rules for electrode arrangement in air-breathing cells were subsequently established by Thorson et al. [64].

Recently, several experimental contributions have demonstrated advances in terms of reactant crossover mitigation and mass transport enhancement by utilizing nonuniform cross-sectional channel geometries [65–67]. For instance, Lopez-Montesinos et al. [65] introduced a microfluidic fuel cell with a bridge-shaped microchannel designed to confine the diffusive liquid–liquid interface away from the electrode areas and to minimize reactant crossover. The bridge-shaped channel is essentially a vertical reorientation of the grooved channel geometry previously developed for gas bubble management in co-laminar flow cells [48]; in the present case, the concept was demonstrated both experimentally and numerically to mitigate crossover while simultaneously enhancing reactant transport to the electrodes. Additionally, Park et al. [67] showed that a microchannel having an H-shaped cross-sectional geometry may increase the current density of the cell while reducing the downstream mixing width when compared to a standard rectangular channel.

A similar approach to reduce diffusive mixing is the use of a porous separator between the co-laminar fuel and oxidant streams, which can facilitate reduced crossover and mixing or enlarged electrochemical chamber dimensions beyond

Fig. 4.7 Chip-embedded thin film current collector designed to reduce contact resistance in cells with porous electrodes. Reproduced with permission from Lee et al. [70]. Copyright Elsevier (2012)

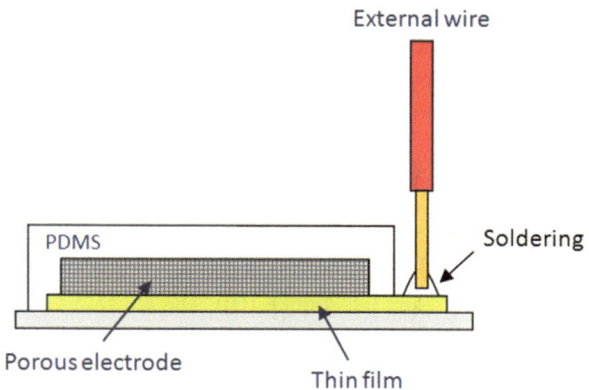

those prescribed by the co-laminar interface alone [55, 64, 68, 69]. Although nonselective porous separators do not suffer from the same cost and durability concerns associated with ionomer membranes, they do serve essentially the same purpose and may add significant complexity and cost to the fabrication process. The porous separator is conceptually similar to the ion-conducting membrane used to separate the half-cells of conventional MEA-based fuel cells, though the porous nature of the separator permits direct liquid–liquid contact between the two streams and therefore (at least partially) qualifies as "membraneless." Nevertheless, porous separators are conveniently able to retain sufficient liquid–liquid contact at the co-laminar flow interface to provide good ionic conduction between the electrodes. The mass transport benefits of the staggered herringbone pattern were strategically exploited by Da Mota et al. [68] by integrating a porous separator at the center of the microchannel to reduce the otherwise chaotic mixing of the two streams. The combination of herringbone ridges and convective barrier was shown to result in high power densities up to $270\,mW\,cm^{-2}$ with sodium borohydride and cerium ammonium nitrate as fuel and oxidant. Alkaline and acidic mixed media conditions were applied to additionally boost the cell voltage and kinetics of the respective half-cells.

With elegant solutions for enhanced mass transport rates and shifting of transport limitations into high current density operation come new challenges and limitations on performance. Most notably, the performance of cells utilizing porous electrodes has been shown to be largely constrained by ohmic resistance and associated voltage losses. A substantial contact resistance between the external wires and porous electrodes is anticipated owing to the highly porous nature of the electrodes. Lee et al. investigated the overall ohmic resistance of the microfluidic fuel cell with flow-through porous electrodes [57] and proposed to incorporate a current collector to reduce the contact resistances [70]. Inspired by the bipolar plates in conventional MEA-based fuel cells, the custom-designed thin-film current collector shown schematically in Fig. 4.7 provided direct physical contact with the porous electrodes and largely increased net contact area. A thin film of Au was chosen as the primary material due to its proven compatibility with acidic environments. In addition, the micromachining-based thin film process was compatible with the overall cell fabrication procedure without major modifications. The prototype microfluidic fuel cell with chip-embedded current collectors demonstrated a 79 % increase in peak power density compared to an otherwise similar cell and achieved a class-leading volumetric peak power density of 6.2 W cm^{-3}.

While reduced microchannel dimensions are known to provide increased surface area and enhanced mass transport rates, the size of the center channel employed in microfluidic fuel cells is dictated by the requirements for reactant crossover prevention, resulting in optimum channel dimensions on the order of hundreds of micrometers to millimeters (cf., Eq. (2.10)). The advantages of submicron length scales can however be uniquely exploited by integration of nanoporous electrodes with the flow-through cell architecture. This arrangement enables reactant flow in nanoscale conduits inside the electrodes without compromising the stability of the co-laminar flow in the microchannel at the center of the device. The first nanofluidic fuel cell designed to capture this opportunity was recently demonstrated by Lee and Kjeang [71]. The proposed cell featured carbon nanofoam electrodes with 10–100 nm characteristic pore size, three orders of magnitude smaller than for the previously employed carbon paper electrodes [57] with a corresponding three orders of magnitude higher active surface area. The nanofluidic fuel cell prototype was found to substantially increase performance in the low current density regime through enhanced kinetics and surface area although the expected mass transport benefits in the high current density regime were not achieved due to the relatively high ohmic resistance of the nanofoam material.

4.6 Microfluidic Redox Batteries

The technical development of membraneless microfluidic redox batteries poses additional research challenges than those elaborated on previously in this chapter. In principle, a microfluidic fuel cell can be considered a primary redox battery operating in discharge mode. However, a rechargeable secondary redox battery must be capable of efficient operation in both discharge and charge modes using a closed system of reactants. Proven redox chemistries previously established for large-scale redox flow batteries are useful for development of microfluidic battery architectures. The all-vanadium redox chemistry is a particularly convenient choice considering that cross-contamination is not an issue and mixed reactants and products may be regenerated directly for reuse. With previous microfluidic fuel cell designs this regeneration would normally be performed off-chip using a separate charging device. Notably however, the structure of the flow-through cell enables in situ regeneration. Specifically, proof-of-concept in situ regeneration of the initial fuel and oxidant species was established by operating the fuel cell in the reverse direction with applied power [57]. As illustrated in Fig. 4.4d, the regeneration of reactants in reverse mode is evidenced by the color changes of the charged reactants emerging from the porous electrodes into the supply channels.

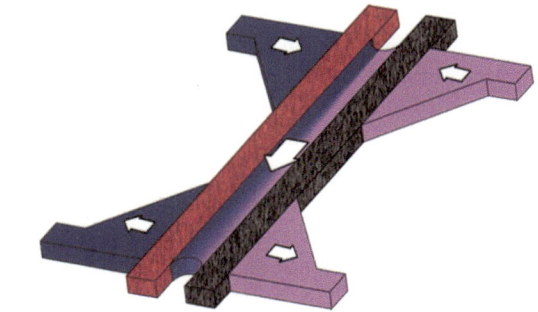

Fig. 4.8 Symmetric and monolithic microfluidic redox battery architecture with flow-through porous electrodes and full recirculation and regeneration capabilities [73]

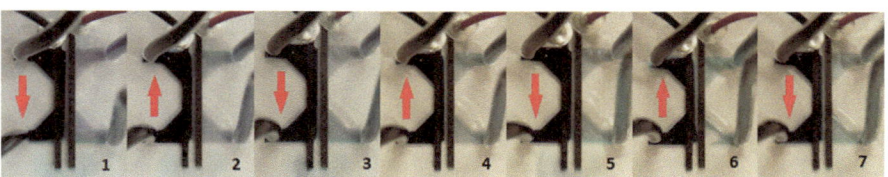

Fig. 4.9 Stepwise forward/reverse recirculation of vanadium redox electrolytes in the symmetric microfluidic redox battery architecture (cf., Fig. 4.8) operated in discharge mode [73]

Until very recently, nearly all of the research on microfluidic electrochemical cells has focused on single-pass fuel cell operation. Most cells were designed with a single outlet leading to the unavoidable mixing of unused reactants and thereby preventing further recirculation. When operated below 100 % fuel utilization conditions, this limits the overall energy efficiency of the cells. A symmetric dual-pass flow-through architecture was recently proposed as a potential solution for reactant recirculation in microfluidic cells [72]. In the first generation of this device [73], as shown in Fig. 4.8, the reactants enter the cell through the two inlets, pass through the first electrode section, flow downstream in the co-laminar center channel, and pass through the electrodes a second time before exiting the cell via the two outlets. The flow can then be reversed to recirculate the same reactants back through the cell and thereby reversing the position of the inlets and outlets. The experiment conducted in Fig. 4.9 depicts seven consecutive forward/reverse runs, with vanadium redox electrolytes being discharged at 1.2 V. The changing colors after each run indicate the gradual decrease of reactant concentration. As an additional benefit of the present cell design, the increased convective velocity caused by the dual-pass configuration improved the mass transport rates and resulted in higher power density at a given flow rate [73].

In situ regenerative operation, as originally demonstrated by Kjeang et al. [57], was uniquely exploited by the symmetric dual outlet architecture of this device to complete a full charge–discharge cycle [72]. The implications of reactant crossover for regeneration are more severe than for single-pass discharge operation and must be carefully addressed through cell design. The proposed microfluidic redox battery (MRB), which is a derivative of the symmetric dual-pass cell design shown in Fig. 4.8, was custom designed to minimize crossover diffusion at the splitting point of the streams. With these design improvements, the cell demonstrated the capability to serve as a rechargeable redox flow battery with ~20 % full cycle energy efficiency [72]. Later in the same year, a rechargeable microfluidic battery operating on hydrogen and bromine was developed by Braff and coworkers at Massachusetts Institute of Technology and published in *Nature Communications* [74]. This membraneless battery combines a liquid phase bromine/hydrobromic acid redox cathode with a gaseous hydrogen anode of the gas diffusion electrode type previously applied for air-breathing cathodes. The rapid kinetics of both half-cells resulted in record breaking power density and round trip voltage efficiency up to 90 %, which represents a significant milestone in this field and further highlights the usefulness of membraneless microfluidic cell architectures for both fuel cells and flow batteries.

4.7 Cell Arrays and Stacks

As the technology is emerging, the majority of microfluidic fuel cell and battery devices reported to date have been proof-of-concept single cells. The voltage and overall power output of these devices were generally less than 1 V and 10 mW, respectively, which is sufficient for research and demonstration purposes but generally inadequate for real-world products and applications. Therefore, scale-up or integration of multiplexed cells is critical to the application of microfluidic fuel cell and battery technology. Several scale-up methodologies have been reported to date for microfluidic fuel cells. Ferrigno et al. [4] demonstrated a planar monolithic array of three cells in a single chip with separate inlets and outlets for each cell. The unit cells were based on the original Y-shaped channel geometry with horizontal co-laminar streaming and electrodes positioned on the bottom wall (Fig. 1.1a). As intended by design, the array generated approximately three times the power of the unit cell, and when connected in series, reached operational voltages up to 2.4 V that are practical for electronic applications. While planar arrays of this type are convenient from a fabrication perspective, they require substantial "overhead" volume of passive materials that confine the volumetric power density of the device. Cohen et al. [7, 12] reported a more compact planar expansion methodology for parallel microchannels with combined inlets and outlets. The prototype multichannel cell employed five parallel rectangular microchannels of the F-shaped design (Fig. 1.1c). The flow was sufficiently uniform to reach a power output that scaled linearly with the results obtained for a single microchannel [7]. This planar design facilitates custom-fabricated channels with dimensions tuned for specific power requirements. Moreover, the expansion of a single microchannel was successfully demonstrated up to 5 mm in width for a 5 cm long channel, with verified linear scaling of power output. The potential for vertical stacking was also confirmed in this work.

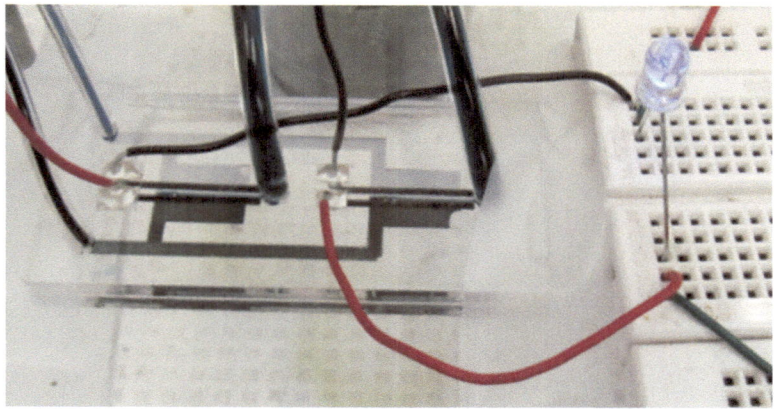

Fig. 4.10 Image of a unilateral two-cell array powering an LED. Reproduced with permission from Ho and Kjeang [75]. Copyright ASME (2013)

Two microchannels placed on top of each other and separated by electrodes were shown to produce twice the power of a single channel without increasing the total volume of the device. While these scale-up methods are promising, the overall power output of the devices was only about 1 mW or less, restricted by transport and solubility of dissolved oxygen, and the fuel utilization was relatively low.

Two planar cell multiplexing strategies were reported by Ho and Kjeang [75], featuring a nonsymmetric unilateral design and a symmetric bilateral device architecture, both of which employed two cells with shared fluidic inlet ports and flow-through porous electrodes. The normalized performance obtained with the two prototype array cells was found to be equivalent to previously reported data for single cells, in this case doubling the device level voltage and power output, thereby demonstrating the feasibility of both expansion concepts. As shown in Fig. 4.10, the unilateral two-cell array device was able to power an LED by directly connecting its terminals to the load, which is normally not possible with a single-cell device due to inadequate cell voltage. Salloum and Posner [76] proposed a third electrolyte stream based on the crossover mitigation concept previously published by Sun et al. [51] to permit reutilization of unused fuel by a second cell positioned in tandem, downstream from the first cell [76]. This strategy was found to be effective in order to improve overall fuel utilization and increase total power output of the device, although the performance of the downstream cell was lower than that of the upstream cell due to partially depleted reactants and the third electrolyte stream added complexity in terms of system design.

In contrast to regular planar electrodes, the geometry and mechanical properties of rod-shaped electrodes enable unique three-dimensional microfluidic cell architectures. An array architecture fuel cell was developed by Kjeang et al. [55] based on a hexagonal array of graphite rods mounted in a single cavity, as depicted in Fig. 4.11. In this case, the flow area between the rods exhibited microfluidic laminar flow characteristics similar to those of a planar unit cell. The array cell had 12

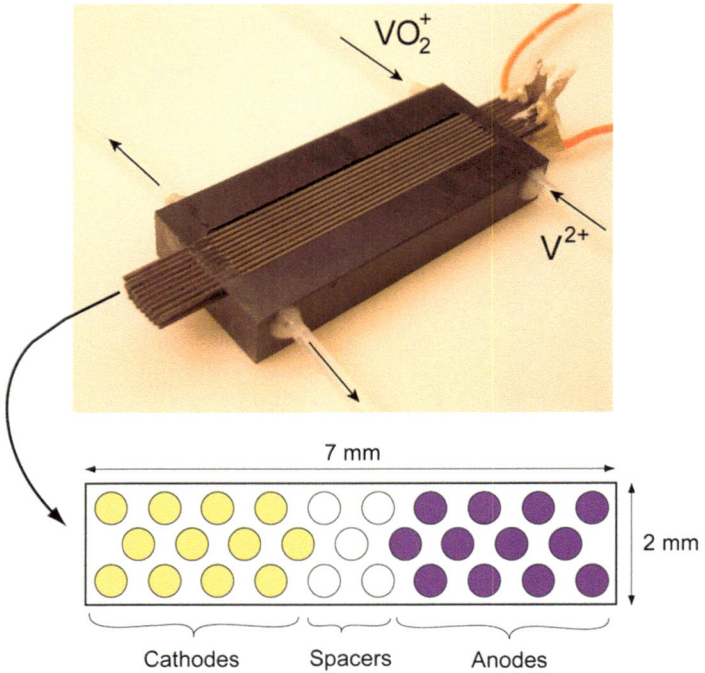

7 mm

2 mm

Cathodes   Spacers   Anodes

Fig. 4.11 A microfluidic fuel cell expanded in three dimensions using a hexagonal array of rod electrodes. Reproduced with permission from Kjeang et al. [55]. Copyright Elsevier (2007)

anodes and 12 cathodes, and the 5 rods in the center were electrically insulated to compensate for the co-laminar interdiffusion zone. When operated on vanadium redox electrolyte at a given flow rate, the array cell produced an order of magnitude more power than a planar unit cell. Specifically, power and current levels of 28 mW and 86 mA were demonstrated, and the fuel utilization was significantly higher than for the planar unit cell. The array cell configuration may be readily expanded in both vertical (preferable) and horizontal directions to increase capacity. Scale-up requires only an enlarged cavity, in contrast to the volumetric costs of stacking of planar cells. Zhu et al. [77] further explored this concept by uniquely pairing the cylindrical array electrode configuration with an air-breathing gas diffusion cathode. This device operated on formic acid fuel showed practical performance levels with relatively high fuel utilization up to 88 % at the lowest flow rate and was able to accommodate CO_2 gas evolution from the anode by bubble capturing and removal in the space between the cylindrical rod electrodes.

Minimizing the volume of supporting materials used for encapsulation, sealing, and clamping is an important consideration for scale-up of microfluidic fuel cells without loss of overall power density when normalized by the whole size of the device. The all-polymer microfluidic fuel cell fabrication scheme developed by Hollinger et al. [78] is capable of producing approximately 1 mm thick cells that are amenable to multiplexing and stacking in both horizontal and vertical directions.

Inspired by conventional MEA technology, these cells utilized thin Kapton films as bonded electrode frames. A vertical plate-frame microfluidic fuel cell stack for cells with flow-through porous electrodes was demonstrated by Moore et al. [79] that leveraged the same general strategy, showing the benefit of increased total power output although the power density was reduced when compared to previously published single cells [57].

A prototype system-integrated microfluidic fuel cell stack based on the air-breathing direct methanol laminar flow fuel cell technology [42] has been reported by INI Power Systems. A combination of planar and vertical stacking methods was employed to scale the system and increase its power output. With respect to fuel utilization, a fuel and electrolyte separation and recirculation system was proposed at the cost of added complexity and reduced energy density of the complete fuel cell system.

References

1. W. Sung, J.-W. Choi, J. Power. Sources **172**, 198–208 (2007)
2. M. Togo, A. Takamura, T. Asai, H. Kaji, M. Nishizawa, Electrochim. Acta **52**, 4669–4674 (2007)
3. C.M. Moore, S.D. Minteer, R.S. Martin, Lab Chip **5**, 218–225 (2005)
4. R. Ferrigno, A.D. Stroock, T.D. Clark, M. Mayer, G.M. Whitesides, J. Am. Chem. Soc. **124**, 12930–12931 (2002)
5. E.R. Choban, L.J. Markoski, A. Wieckowski, P.J.A. Kenis, J. Power. Sources **128**, 54–60 (2004)
6. E.R. Choban, P. Waszczuk, P.J.A. Kenis, Electrochem. Solid State Lett. **8**, A348–A352 (2005)
7. J.L. Cohen, D.A. Westly, A. Pechenik, H.D. Abruna, J. Power. Sources **139**, 96–105 (2005)
8. F.M. Cuevas-Muniz, M. Guerra-Balcazar, J.P. Esquivel et al., J. Power. Sources **216**, 297–303 (2012)
9. A. Dector et al., Int. J. Hydrogen Energ. **38**, 12617–12622 (2013)
10. D. Morales-Acosta, H. Rodríguez, L.A. Godinez, L.G. Arriaga, J. Power. Sources **195**, 1862–1865 (2010)
11. D. Morales-Acosta, M.D. Morales-Acosta, L.A. Godinez, L. Álvarez-Contreras, S.M. Duron-Torres, J. Ledesma-García, L.G. Arriaga, J. Power. Sources **196**, 9270–9275 (2011)
12. J.L. Cohen, D.J. Volpe, D.A. Westly, A. Pechenik, H.D. Abruna, Langmuir **21**, 3544–3550 (2005)
13. E.R. Choban, J.S. Spendelow, L. Gancs, A. Wieckowski, P.J.A. Kenis, Electrochim. Acta **50**, 5390–5398 (2005)
14. S. Hasegawa, K. Shimotani, K. Kishi, H. Watanabe, Electrochem. Solid State Lett. **8**, A119–A121 (2005)
15. F.R. Brushett, R.S. Jayashree, W.-P. Zhou, P.J.A. Kenis, Electrochim. Acta **54**, 7099–7105 (2009)
16. R.A. Bullen, T.C. Arnot, J.B. Lakeman, F.C. Walsh, Biosens. Bioelectron. **21**, 2015–2045 (2006)
17. E. Katz, A.N. Shipway, I. Willner, Biochemical fuel cells, in *Handbook of Fuel Cells—Fundamentals, Technology and Applications*, ed. by W. Vielstich, H.A. Gasteiger, A. Lamm, vol. 1 (Wiley, New York, 2003)
18. E. Kjeang, N. Djilali, D. Sinton, J. Power. Sources **186**, 353–369 (2009)
19. I. Willner, E. Katz, Angew. Chem. Int. Ed. **39**, 1180–1218 (2000)
20. M.N. Arechederra et al., Electrochim. Acta **56**, 1585–1590 (2011)
21. J.W. Lee, E. Kjeang, Biomicrofluidics **4**, 041301 (2010)
22. N.L. Akers, C.M. Moore, S.D. Minteer, Electrochim. Acta **50**, 2521–2525 (2005)
23. S. Topcagic, S.D. Minteer, Electrochim. Acta **51**, 2168–2172 (2006)
24. M. Togo, A. Takamura, T. Asai, H. Kaji, M. Nishizawa, J. Power. Sources **178**, 53–58 (2008)

25. K.G. Lim, G.T.R. Palmore, Biosens. Bioelectron. **22**, 941–947 (2007)
26. A. Zebda, L. Renaud, M. Cretin, C. Innocent, R. Ferrigno, S. Tingry, Sens. Actuators B **149**, 44–50 (2010)
27. M.J. Gonzales-Guerrero et al., Lab Chip **13**, 2972 (2013)
28. D. Ye et al., Int. J. Hydrogen Energ. **38**, 15710–15715 (2013)
29. R.S. Jayashree, L. Gancs, E.R. Choban, A. Primak, D. Natarajan, L.J. Markoski, P.J.A. Kenis, J. Am. Chem. Soc. **127**, 16758–16759 (2005)
30. C. Chen, P. Yang, J. Power. Sources **123**, 37–42 (2003)
31. G. Girishkumar, B. McCloskey, A.C. Luntz, S. Swanson, W. Wilcke, J. Phys. Chem. Lett. **1**, 2193–2203 (2010)
32. S.A. Mousavi Shaegh, N.-T. Nguyen, S.H. Chan, J. Micromech. Microeng. **20**, 105008 (2010)
33. S.A. Mousavi Shaegh, N.-T. Nguyen, S.H. Chan, W. Zhou, Int. J. Hydrogen Energ. **37**, 3466–3476 (2012)
34. R.S. Jayashree, S.K. Yoon, F.R. Brushett, P.O. Lopez-Montesinos, D. Natarajan, L.J. Markoski, P.J.A. Kenis, J. Power. Sources **195**, 3569–3578 (2010)
35. R.S. Jayashree, D. Egas, J.S. Spendelow, D. Natarajan, L.J. Markoski, P.J.A. Kenis, Electrochem. Solid State Lett. **9**, A252–A256 (2006)
36. S.A. Mousavi Shaegh, N.-T. Nguyen, S.H. Chan, J. Power. Sources **209**, 312–317 (2012)
37. D.T. Whipple, R.S. Jayashree, D. Egas, N. Alonso-Vante, P.J.A. Kenis, Electrochim. Acta **54**, 4384–4388 (2009)
38. J. Ma et al., J. Electrochem. Soc. **160**, F859–F866 (2013)
39. S. Tominaka, S. Ohta, H. Obata, T. Momma, T. Osaka, J. Am. Chem. Soc. **130**, 10456–10457 (2008)
40. S. Tominaka et al., Energ. Environ. Sci. **2**(10), 1074–1077 (2009)
41. S. Tominaka et al., Energ. Environ. Sci. **2**(8), 849–852 (2009)
42. L.J. Markoski, *The laminar flow fuel cell: a portable power solution*, in *8th Annual International Symposium Small Fuel Cells 2006; Small Fuel Cells for Portable Applications,* The Knowledge Foundation, Washington, DC, 2006
43. C.P. de Leon, F.C. Walsh, A. Rose, J.B. Lakeman, D.J. Browning, R.W. Reeve, J. Power. Sources **164**, 441–448 (2007)
44. G.H. Miley, N. Luo, J. Mather, R. Burton, G. Hawkins, L.F. Gu, E. Byrd, R. Gimlin, P.J. Shrestha, G. Benavides, J. Laystrom, D. Carroll, J. Power. Sources **165**, 509–516 (2007)
45. N.A. Choudhury, R.K. Raman, S. Sampath, A.K. Shukla, J. Power. Sources **143**, 1–8 (2005)
46. A. Li, S.H. Chan, N.T. Nguyen, J. Micromech. Microeng. **17**, 1107–1113 (2007)
47. J.-C. Shyu, C.-S. Wei, C.-J. Lee, C.-C. Wang, Appl. Therm. Eng. **30**, 1863–1871 (2010)
48. E. Kjeang, A.G. Brolo, D.A. Harrington, N. Djilali, D. Sinton, J. Electrochem. Soc. **154**, B1220–B1226 (2007)
49. J.I. Hur, D.D. Meng, C.-J. Kim, J. Microelectromech. Syst. **21**, 476–483 (2012)
50. R. Aogaki, E. Ito, M. Ogata, J. Solid State Electrochem. **11**, 757–762 (2007)
51. M.H. Sun, G.V. Casquillas, S.S. Guo, J. Shi, H. Ji, Q. Ouyang, Y. Chen, Microelectron. Eng. **84**, 1182–1185 (2007)
52. L. Gu, N. Luo, G.H. Miley, J. Power. Sources **173**, 77–85 (2007)
53. C.P. de Leon, A. Frias-Ferrer, J. Gonzalez-Garcia, D.A. Szanto, F.C. Walsh, J. Power. Sources **160**, 716–732 (2006)
54. W. SkyllasKazacos, C. Menictas, M. Kazacos, J. Electrochem. Soc. **143**, L86–L8856 (1996)
55. E. Kjeang, J. McKechnie, D. Sinton, N. Djilali, J. Power. Sources **168**, 379–390 (2007)
56. E. Kjeang, B.T. Proctor, A.G. Brolo, D.A. Harrington, N. Djilali, D. Sinton, Electrochim. Acta **52**, 4942–4946 (2007)
57. E. Kjeang, R. Michel, D.A. Harrington, N. Djilali, D. Sinton, J. Am. Chem. Soc. **130**, 4000–4006 (2008)
58. E. Kjeang, R. Michel, D. Sinton, N. Djilali, D.A. Harrington, Electrochim. Acta **54**, 698–705 (2008)
59. K.S. Salloum, J.R. Hayes, C.A. Friesen, J.D. Posner, J. Power. Sources **180**, 243–252 (2008)

60. S.K. Yoon, G.W. Fichtl, P.J.A. Kenis, Lab Chip **6**, 1516–1524 (2006)
61. A.D. Stroock, S.K.W. Derringer, A. Ajdari, I. Mezić, H.A. Stone, G.M. Whitesides, Science **295**, 647 (2002)
62. J.D. Kirtland, G.J. McGraw, A.D. Stroock, Phys. Fluids **18**, 073602 (2006)
63. J. Lee, K.G. Lim, G.T.R. Palmore, A. Tripathi, Anal. Chem. **79**, 7301–7307 (2007)
64. M.R. Thorson, F.R. Brushett, C.J. Timberg, P.J.A. Kenis, J. Power. Sources **218**, 28–33 (2012)
65. P.O. López-Montesinos, N. Yossakda, A. Schmidt, F.R. Brushett, W.E. Pelton, P.J.A. Kenis, J. Power. Sources **196**, 4638–4645 (2011)
66. W. Huo, Y. Zho, H. Zhang, Int. J. Electrochem. Sci. **8**, 4827–4838 (2013)
67. H.B. Park, K.H. Lee, H.J. Sung, J. Power. Sources **226**, 266–271 (2013)
68. N. Da Mota, D.A. Finkelstein, J.D. Kirtland, C.A. Rodriguez, A.D. Stroock, H.D. Abruña, J. Am. Chem. Soc. **134**, 6076–6079 (2012)
69. A.S. Hollinger, R.J. Maloney, R.S. Jayashree, D. Natarajan, L.J. Markoski, P.J.A. Kenis, J. Power. Sources **195**, 3523–3528 (2010)
70. J.W. Lee, E. Kjeang, Int. J. Hydrogen Energ. **37**, 9359–9367 (2012)
71. J.W. Lee, E. Kjeang, J. Power. Sources **242**, 472–477 (2013)
72. J. Lee, M.-A. Goulet, E. Kjeang, Lab Chip **13**, 2504–2507 (2013)
73. M.-A. Goulet, E. Kjeang, Electrochim. Acta (2014), in press
74. W.A. Braff et al., Nat. Commun. **4**, 2346 (2013)
75. B. Ho, E. Kjeang, J. Fluid. Eng. **135**, 021304 (2013)
76. K.S. Salloum, J.D. Posner, J. Power. Sources **196**, 1229–1234 (2011)
77. X. Zhu et al., J. Power. Sources **247**, 346–353 (2014)
78. A.S. Hollinger, P.J.A. Kenis, J. Power. Sources **240**, 486–493 (2013)
79. S. Moore, D. Sinton, D. Erickson, J. Power. Sources **196**, 9481–9487 (2011)

Chapter 5
Modeling

The operation of microfluidic fuel cells and batteries is conceptually governed by conservation of mass, momentum, species, and charge. The geometrical complexity and significant coupling of different physics generally require computational approaches to solve the complete set of governing equations. Computational fluid dynamics (CFD) tools with multiphysics capabilities are therefore essential in the model development. In contrast to macroscale fluid mechanics where turbulence is a major challenge, the main challenges in CFD modeling of microfluidic devices are in the application of appropriate boundary conditions and in modeling species transport. The addition of electrochemical reaction kinetics in microfluidic devices substantially increases the modeling complexity by the coupling with mass/species/charge transport and fluid flow via the source/sink terms of the reactions. In the context of microfluidic fuel cells, this area was first investigated by Bazylak et al. [1]. A 3-D CFD framework coupled with convective/diffusive mass transport (infinite dilution) and electrochemical reaction rate models for both anode and cathode was employed to analyze the original T-shaped formic acid/dissolved oxygen microfluidic fuel cell with horizontal co-laminar streaming. Various cross-sectional channel geometries and electrode configurations were considered, targeting enhanced fuel utilization while minimizing fuel/oxidant mixing. A high aspect ratio (width/height) channel geometry with electrodes placed on the top and bottom walls was found to enable significantly improved fuel utilization and reduced mixing width. As illustrated in Fig. 5.1, the numerical study also suggested the implementation of a tapered electrode design that accommodates the growth of the co-laminar mixing zone in the downstream direction.

Chang et al. [2, 3] provided an extended model with Butler–Volmer electrochemical reaction kinetics and the capability of predicting complete polarization curves. The results obtained for Y-shaped [2] and F-shaped [3] formic acid/dissolved oxygen-based cells were in good agreement with previous experimental studies [4, 5] and confirmed the cathodic activity and mass transport limitation of these cells. Consequently, the predicted cell performance was essentially independent of anodic formic acid concentration. The numerical results also recommended high aspect

E. Kjeang, *Microfluidic Fuel Cells and Batteries*, SpringerBriefs in Energy,
DOI 10.1007/978-3-319-06346-1_5, © The Author(s) - SpringerBriefs 2014

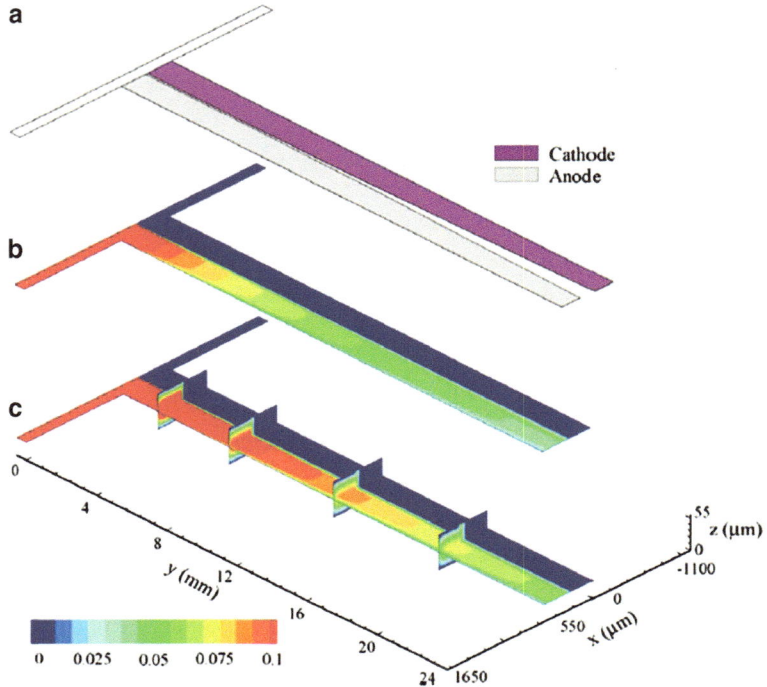

Fig. 5.1 Computational modeling results for a standard T-shaped microfluidic fuel cell with tapered electrodes on top and bottom surfaces, showing (**a**) the tapered electrode geometry and (**b**, **c**) fuel concentration contours in the center plane with vertical projections in (**c**). Reproduced with permission from Bazylak et al. [1]. Copyright Elsevier (2005)

ratio channel geometry, high Péclet number, high oxygen concentration, and a thick cathode catalyst layer to improve the performance. This work was later extended by a 2-D theoretical model of the cathode kinetics under co-laminar flow [6]. A Butler–Volmer model [7] was also developed for a microfluidic fuel cell using hydrogen peroxide as both fuel and oxidant in mixed media conditions [8] and applied to investigate the effects of species transport and geometrical design. The simulated fuel cell performance results were invariant at flow rates above 0.1 mL min^{-1}, indicating the absence of the commonly encountered cathodic transport limitation attributed to the liquid oxidant. Two-phase flow and transport effects related to oxygen gas evolution from hydrogen peroxide decomposition were not considered. However, it was found that increasing the surface area and thickness of the catalyst layers can enhance current density.

A pioneering modeling contribution in the microfluidic biofuel cell domain focused on current extraction from a sequence of consecutive biocatalyzed reactions [9]. To harness additional current from consecutive reactions and thereby improve fuel utilization, strategic patterning of multiple enzyme electrodes is required. For instance, a cascade of enzymes could be patterned in separate patches on each electrode of the fuel cell, as illustrated in Fig. 5.2. This opportunity was investigated

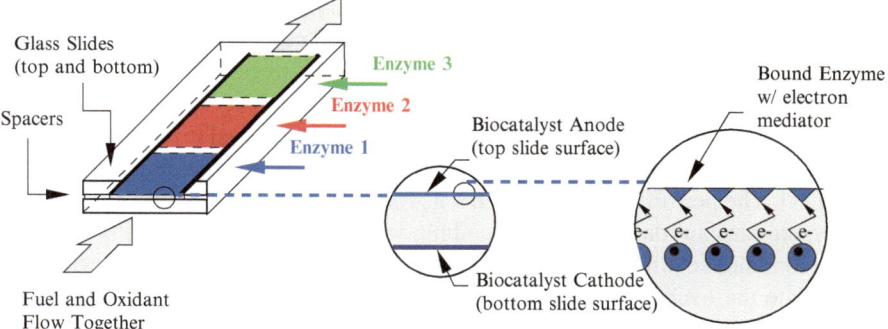

Fig. 5.2 Microfluidic biofuel cell with consecutive reactions catalyzed by a sequence of different enzymes. Reproduced with permission from [9]. Copyright Elsevier (2005)

through numerical simulations using a generic computational model of species transport in microchannels with heterogeneous electrochemical reactions and Michaelis–Menten enzyme kinetics as boundary conditions [9]. This first computational study of microfluidic biofuel cell technology provided guidelines for the design and fabrication of microfluidic biofuel cells exploiting consecutive reactions. Separated and mixed enzyme patterns in different proportions were analyzed for various Péclet numbers. The mixed transport regime, at medium *Pe*, was shown to be particularly attractive while current densities were maintained close to maximum levels. Mixed enzyme patterning tailored with respect to individual turnover rates was found to enable high current densities combined with nearly complete fuel utilization and provide the best overall performance.

The development of microfluidic electrochemical cells stands to benefit significantly from computational and theoretical models that are capable of predicting the performance of different cell designs and can be employed to guide prototyping of new devices. Among the dozens of modeling studies published to date, the majority have considered the standard Y-junction geometry with planar flow-by electrodes (Fig. 1.1a, b). As described above, the first contributions focused on mass transport in the depletion boundary layer which forms at the electrode surface in the case of slow cross-stream diffusive transport in the laminar flow combined with relatively rapid electrochemical kinetics [1, 2, 7, 10, 11]. A modeling contribution by Yoon et al. [12] suggested the use of multiple inlets or outlets to replenish the reactants. Others have explored strategies to minimize diffusion-based crossover of reactant species with the most common suggestions being to taper the electrodes away from the co-laminar interface [1] or to taper the channel downstream to increase the convective velocity of the electrolyte [12, 13]. Modeling predictions have also suggested the viability of concentric laminar flow streaming in novel cylindrical and star-shaped cell architectures with improved performance versus conventional horizontal or vertical co-laminar flows [14]. Regardless of the geometry chosen, many of the recommendations emerging from modeling results are yet to be implemented into a working device and verified experimentally.

More recent modeling contributions have analyzed the performance of the air-breathing cell design [15–20]. Interestingly, Xuan et al. [16] demonstrated that even air-breathing cells will eventually be limited by oxygen starvation when the anode performance exceeds 200 mA cm^{-2}. A numerical modeling scheme was also developed for the monolithic air-breathing microfluidic fuel cell design in order to provide improved fundamental understanding of its unique operational characteristics [21]. The performance of the cell was found to be predominantly constrained by oxygen transport due to cathode flooding, which can be mitigated using a hydrophobic ionomer coating. A system-level energy and exergy analysis was performed to simulate the overall effectiveness of the microfluidic cell design under various operating conditions [22]. It was determined that fuel recirculation can raise the exergy efficiency of the device operation when the efficiency of the micropump employed for this purpose is sufficiently high.

Due to the increased modeling complexity, only two studies on microfluidic cells with flow-through porous electrodes were published to date [23, 24]. The principal challenge lies in the local coupling of Darcy flow, electrochemical kinetics, and convective/diffusive mass transport inside the nonuniform porous electrodes. Krishnamurthy et al. [23] resolved this issue by applying surface concentration dependent Butler–Volmer kinetics combined with an empirical mass transport relationship for flow over a single fiber. The accuracy of this model was later improved by targeted measurements of kinetic parameters for porous carbon electrodes [25], shown to differ considerably from the equivalent parameters historically measured for planar electrodes. The recent modeling study by Sprague and Dutta [24] suggested the use of nanoporous electrodes to specifically increase the advective flux within the electrical double layer (EDL). A laminar flow model was formulated based on the Poisson–Nernst–Planck and Frumkin–Butler–Volmer equations to simulate advective transport and electrochemical reactions in nanochannels. With pore sizes on the order of the EDL width, several interesting phenomena may occur in the presence of laminar flow. For instance, double layer overlap may influence the structure of the electrode–electrolyte interface and create an upstream region of zero charge at the electrodes. Perhaps more importantly, electrolyte advection within the EDL was stipulated to enhance the kinetic performance of the electrodes, which is in agreement with the experimental results obtained with the nanofluidic fuel cell prototype demonstrated by Lee and Kjeang [26].

The first modeling works on rechargeable microfluidic redox batteries were recently contributed by Braff et al. [27, 28]. Numerical and analytical models were developed and validated for the membraneless hydrogen bromine laminar flow battery, comprising sufficient theory and empirically measured parameters to describe both charging and discharging reactions. A mathematical model based on boundary layer analysis of mass transport was derived to provide analytical approximations for the current–voltage relationship in the limit of large Péclet numbers [28]. Analytical descriptors were made possible in this case by the rapid electrochemical kinetics of both half-cells with essentially negligible activation overpotentials. In general, however, a complete numerical model is required to resolve the coupling between electrochemical kinetics and mass transport for more complex chemistries and/or cell geometries.

References

1. A. Bazylak, D. Sinton, N. Djilali, J. Power. Sources **143**, 57–66 (2005)
2. M.H. Chang, F. Chen, N.S. Fang, J. Power. Sources **159**, 810–816 (2006)
3. F.L. Chen, M.H. Chang, M.K. Lin, Electrochim. Acta **52**, 2506–2514 (2007)
4. J.L. Cohen, D.A. Westly, A. Pechenik, H.D. Abruna, J. Power. Sources **139**, 96–105 (2005)
5. E.R. Choban, L.J. Markoski, A. Wieckowski, P.J.A. Kenis, J. Power. Sources **128**, 54–60 (2004)
6. W.Y. Chen, F.L. Chen, J. Power. Sources **162**, 1137–1146 (2006)
7. F. Chen, M.-H. Chang, C.-W. Hsu, Electrochim. Acta **52**, 7270–7277 (2007)
8. S. Hasegawa, K. Shimotani, K. Kishi, H. Watanabe, Electrochem. Solid State Lett. **8**, A119–A121 (2005)
9. E. Kjeang, D. Sinton, D.A. Harrington, J. Power. Sources **158**, 1–12 (2006)
10. J. Lee, K.G. Lim, G.T.R. Palmore, A. Tripathi, Anal. Chem. **79**, 7301–7307 (2007)
11. J. Phirani, S. Basu, J. Power. Sources **175**, 261–265 (2008)
12. S.K. Yoon, G.W. Fichtl, P.J.A. Kenis, Lab Chip **6**, 1516–1524 (2006)
13. R. Hassanshahi, M. Fathipour, Int. J. Adv. Renew. Energ. Res. **1**, 649–654 (2012)
14. R.A. Garcia-Cuevas et al., Int. J. Hydrogen Energ. **38**, 14791–14800 (2013)
15. R.S. Jayashree, S.K. Yoon, F.R. Brushett, P.O. Lopez-Montesinos, D. Natarajan, L.J. Markoski, P.J.A. Kenis, J. Power. Sources **195**, 3569–3578 (2010)
16. J. Xuan, D.Y.C. Leung, H. Wang, M.K.H. Leung, B. Wang, M. Ni, Appl. Energ. **104**, 400–407 (2013)
17. J. Xuan, D.Y.C. Leung, M.K.H. Leung, M. Ni, H. Wang, Int. J. Hydrogen Energ. **36**, 9231–9241 (2011)
18. J. Xuan, D.Y.C. Leung, M.K.H. Leung, H. Wang, M. Ni, J. Power. Sources **196**, 9391–9397 (2011)
19. J. Xuan, H. Wang, D.Y.C. Leung, M.K.H. Leung, H. Xu, L. Zhang, Y. Shen, J. Power. Sources **231**, 1–5 (2013)
20. H. Zhang, J. Xuan, H. Xu, M.K.H. Leung, D.Y.C. Leung, L. Zhang, H. Wang, L. Wang, Appl. Energ. **112**, 1–7 (2013)
21. S. Tominaka et al., Energ. Environ. Sci. **4**, 162–171 (2011)
22. H. Zhang, M.K.H. Leung, J. Xuan, H. Xu, L. Zhang, D.Y.C. Leung, H. Wang, Int. J. Hydrogen Energ. **38**, 6526–6536 (2013)
23. D. Krishnamurthy, E.O. Johansson, J.W. Lee, E. Kjeang, J. Power. Sources **196**, 10019–10031 (2011)
24. I.B. Sprague, P. Dutta, Electrochim. Acta **91**, 20–29 (2013)
25. J.W. Lee et al., Electrochim. Acta **83**, 430–438 (2012)
26. J.W. Lee, E. Kjeang, J. Power. Sources **242**, 472–477 (2013)
27. W.A. Braff et al., Nat. Commun. **4**, 2346 (2013)
28. W.A. Braff et al., J. Electrochem. Soc. **160**, A2056–A2063 (2013)

Chapter 6
Research Trends and Directions

6.1 Publications

The innovative notion of using microscale hydrodynamic engineering in lieu of ionomer membranes is generating an increased interest within the fuel cell and battery research community. The literature citation chart provided in Fig. 6.1a illustrates the rapid growth in annual citations of microfluidic fuel cell publications. The citation trend can be qualitatively compared to the historical data given in Fig. 6.1b for the more established direct methanol fuel cell (DMFC) technology. With the first studies on DMFCs published in the 1980s, the initial growth in citations was rather slow, and it took more than a decade for the annual citations to exceed 500, which was additionally supported by the fuel cell boom at the start of the twenty-first century. In the case of microfluidic fuel cells, the annual citations have grown more rapidly and taken only 8 years to reach the same level of recognition. Notably, the citation growth in several other technological fields including redox flow batteries and alkaline fuel cells (not shown here for brevity) has been even slower than for DMFCs and taken over two decades to generate the 500-mark impact level. While this analysis is only approximate, it reveals that research on microfluidic fuel cells is growing faster than some other electrochemical energy conversion technologies and will likely continue to grow, potentially in a pattern similar to the trend set by DMFCs which recently stabilized around 10,000 citations and over 200 publications per year.

Besides the rapidly increasing citation trend, a detailed review of the actual number of publications reveals more useful information. The publication chart presented in Fig. 6.2 accounts for all peer-reviewed scientific publications that report new findings on co-laminar flow-based microfluidic electrochemical cells, which form the core of this field. To avoid biasing effects, review articles and contributions on peripheral fuel cell technologies with flowing electrolytes [1–4] or mixed reactants [5–12] were not included. Overall, as indicated by the trendline, the total number of scientific articles on microfluidic fuel cells and batteries has increased considerably over the past decade and exceeded 20 per year by the end of 2013.

E. Kjeang, *Microfluidic Fuel Cells and Batteries*, SpringerBriefs in Energy, DOI 10.1007/978-3-319-06346-1_6, © The Author(s) - SpringerBriefs 2014

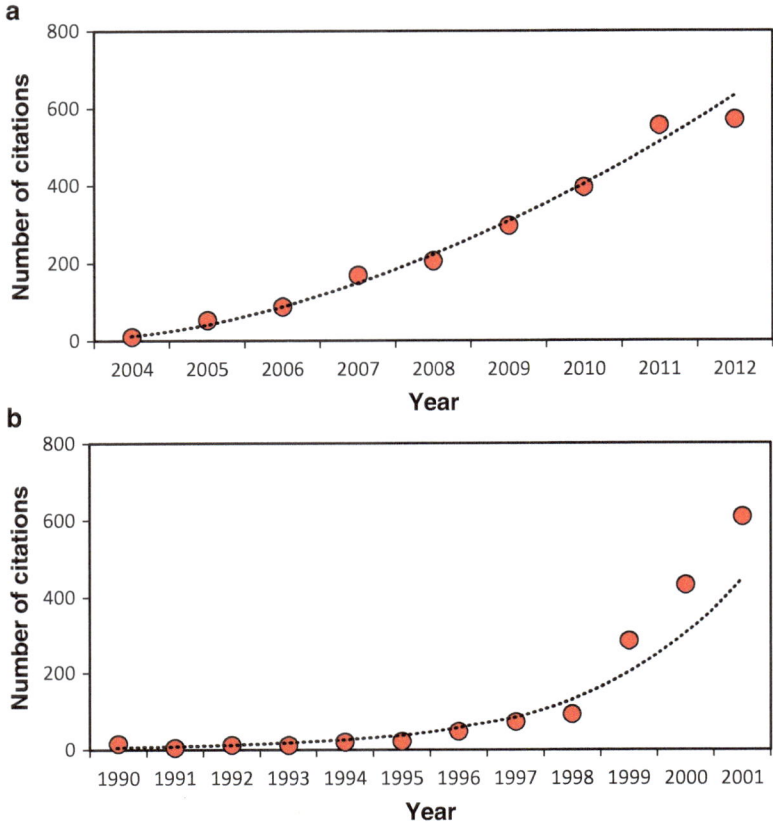

Fig. 6.1 Number of citations per year obtained from Web of Knowledge[SM] for search terms:
(**a**) "microfluidic" + "fuel cell" and (**b**) "direct methanol" + "fuel cell"

It is noteworthy that 2013 was a particularly strong and promising year in terms of
both the number of publications and the overall scope and anticipated impact of
these contributions—again indicating good potential for further growth in citations.
Most of the publications accounted for were contributed by well-established
research groups that were pioneers in this field. However, the present publication
count does not include the numerous theses and conference presentations available
from emerging research groups who are likely to contribute to the future growth of
this field. As indicated in the figure, the total publications to date were dominated
by experimental contributions. The proportion of modeling-focused publications
has also increased in recent years, which has resulted in improved fundamental
understanding and useful guidelines for design of next generation cells. It can be
anticipated that these outcomes may lead to significant near-term growth in experimental prototypes, as has indeed been the case for 2013. However, the majority of
the modeling contributions to date simulated various derivatives of the original

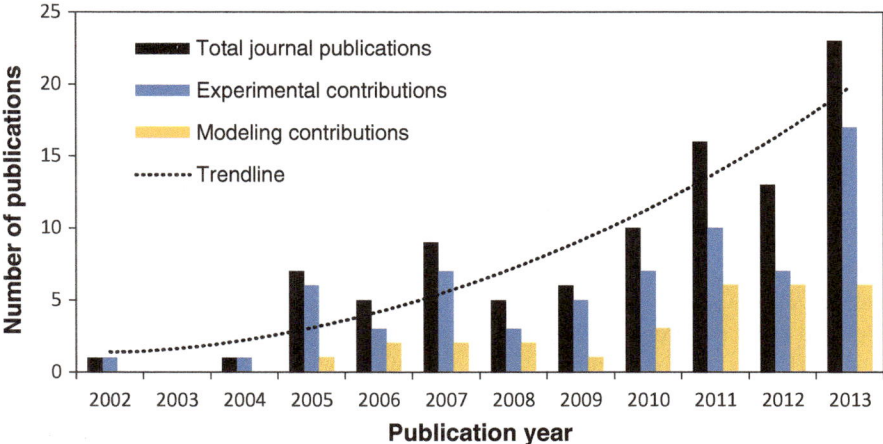

Fig. 6.2 Number of journal publications on co-laminar flow-based microfluidic electrochemical cells (including both fuel cells and batteries). Reproduced and adapted with permission from Goulet and Kjeang [37]

Y-shaped cell with planar flow-by electrode configuration, which may have limited relevance as the field is maturing and progressing towards more advanced cell architectures with high performance. It is also noteworthy that a significant portion of the modeling results and recommendations regarding performance enhancements still lack reliable data for experimental verification. Both of these concerns indicate a gap between the modeling and experimental approaches which needs to be addressed by groups developing both capabilities in-house or seeking more collaboration with complementary groups. Nevertheless, the resurgence of experimental contributions in 2013 shows that the field is advancing in the right direction and holds great promise for future growth.

6.2 Performance

The overall performance of microfluidic fuel cells and batteries has advanced tremendously over the past decade towards levels that appear to be competitive when compared against alternate technologies. However, although common performance measurements can be made on all devices, direct comparisons between different cell architectures are difficult due to several confounding factors. This challenge is addressed here by comparing the leading microfluidic cell devices in each category according to several common and proposed performance metrics. Subsequently, it is discussed why isolated performance comparisons may not holistically reflect the advantages or drawbacks of each cell design, and a few recommendations for future comparative work are offered.

Since the development of the modern membrane electrode assembly (MEA), the most commonly cited performance metrics for traditional fuel cells are the peak power of the entire cell and the peak power density normalized by the geometric electrode area [13]. The power density convention was established to compare single cells with different sizes, provided that current and therefore the power theoretically scales in direct proportion to the electrode surface area. This convention is difficult to adapt to the three-dimensional geometry of microfluidic fuel cells; however, most co-laminar flow cells utilizing planar electrodes continue to follow the power density convention according to the geometric area of a single electrode. In the case of flow-through porous electrodes, performance normalization was initially reported using the projected electrode area, while more recent studies reported densities based on the cross-sectional area normal to the flow [14, 15]. Highly dependent on the aspect ratio of the 3-D porous electrodes, neither choice will give a complete description of the system. Hence, volumetric power density was suggested as a more generally applicable criterion [16].

Fundamentally, both MEA-based and membraneless cells require two electrodes with an ionically conductive electrolyte between them. It is therefore proposed that a volumetric power density normalized by the essential volume of the electrochemical chamber, including both electrodes and the separating electrolyte, would be the most universally applicable metric for these devices. This metric captures any variations in electrolyte channel separation and electrode thickness with the only assumption being that the inlet/outlet flow field manifolds and other structural support elements are comparable between cells. With this new convention, the key microfluidic electrochemical cell technologies with the highest power densities reported to date were converted where possible and presented in Table 6.1. For comparative purposes, estimates for a typical MEA-based vanadium redox battery (VRB) [17, 18] and a DMFC [19] are also included.

Due to the considerable differences in reactant chemistries and testing conditions between the cells presented in Table 6.1, it is difficult to draw final conclusions about which specific device architecture may be optimal. Factors such as reactant species, reactant and electrolyte concentration, flow rate, temperature, separators, patterned electrodes, and the presence of catalysts can all significantly affect the power output of a device. Power density comparisons would ideally be made with devices benchmarked at standardized conditions; for example, a liquid/liquid cell could be benchmarked at room temperature with standardized vanadium electrolytes (e.g., 1 M vanadium in 1 M sulfuric acid) and widely available porous carbon electrodes (e.g., Toray carbon paper). Conducting polarization curve and impedance measurements at a range of flow rates would enable full characterization of fuel utilization, mass transport, and ohmic losses which are inherent to the cell structure, and the peak volumetric power density measurements would then enable a direct comparison with other devices.

Despite the variability in conditions, several useful trends can be extracted from the performance data in Table 6.1. For the cells with flow-through porous electrodes, the performance of the original cell by Kjeang et al. appears to surpass the modified cell by Lee et al. when normalized by projected electrode area.

Table 6.1 Performance data on key microfluidic electrochemical cell technologies

Author (year)	Category	Electrode type	Fabrication	Additional features	Fuel	Oxidant	Peak power (mW)	Peak power density (mW cm^{-2})	Volumetric peak power density (mW cm^{-3})
Ferrigno (2002)	Liquid/liquid	Flow-by	Monolithic	None	1 M V^{2+}	1 M VO$_2^+$	3	38	475
Choban (2005)	Liquid/liquid	Flow-by	Multilayer	Pt catalysts & mixed media	1 M CH$_3$OH	Dissolved O$_2$	3	12	160
Kjeang (2008)	Liquid/liquid	Flow-through	Monolithic	None	2 M V^{2+}	2 M VO$_2^+$	15	121	1,344
Jayashree (2010)	Liquid/gaseous	Flow-by & air-breathing	Multilayer	Pt catalyst	1 M HCOOH	Pure O$_2$	36	55	809
Zebda (2010)	Liquid/liquid	Flow-by	Monolithic	Enzymatic biofuel cell	Glucose	Dissolved O$_2$	0.03	0.55	19
Lee (2012)	Liquid/liquid	Flow-through	Monolithic	Thin-film current collector	2 M V^{2+}	2 M VO$_2^+$	11	93	2,067
Da Mota (2012)	Liquid/liquid	Flow-by	Multilayer	Separator & Pt catalysts & patterned electrodes	0.15 M NaBH$_4$	0.5 M cerium ammonium nitrate (CAN)	135	270	5,190
Braff (2013)	Liquid/gaseous	Flow-by & H$_2$-breathing	Multilayer	Pt catalyst	Pure H$_2$	5 M Br$_2$	199	795	6,625
VRB[a]	Liquid/liquid	Flow-through	MEA	Membrane	1.5 M V^{2+}	1.5 M VO$_2^+$	130,000	150	700
DMFC[b]	Liquid/gaseous	Flow-through & air-breathing	MEA	Membrane & Pt catalyst	3 M CH$_3$OH	Pure O$_2$	450	50	690

Reproduced and adapted with permission from Goulet and Kjeang [37]

[a]VRB MEA dimensions compiled from Skyllas et al. [17] & Zhao et al. [18]. Peak power estimated from upper limits of current density and voltage taken from Zhao et al.

[b]Typical room temperature performance and cell dimensions of DMFC estimated from Xu et al. [19]

When normalized by the cross-sectional electrode area, however, the cell by Kjeang et al. produced 403 mW cm^{-2} while the one by Lee et al. produced a superior 620 mW cm^{-2}. The more universal volumetric power density, which captures all electrode dimensions, accurately verifies this performance gain. With common vanadium reactants, these cells can also be compared to the flow-by cell originally published by Ferrigno et al. Though the concentrations and flow rates differ, the power output of the flow-through cells is significantly larger than for the flow-by cell, likely due to the enhanced mass transport characteristics. This result was unambiguously verified in the 2008 study, in which the same electrolyte species were forced to flow either over or through the same porous carbon material, demonstrating that flow-through porous electrodes lead to significant performance gains across a wide range of flow rates [16]. It is also noteworthy that with similar electrode materials and reactant concentrations, the normalized performance of microfluidic fuel cells with flow-through electrodes compares quite favorably with and even exceeds that of existing commercial VRB technology. With single-cell power output in the 10 mW range, however, the targeted application for microfluidic cells will be quite different to the high-power conventional VRBs with a typical 14-cell stack producing on the order of 1 kW [18]. In summary, recent advances in microfluidic cell technology have generated performance levels comparable to more well-established MEA-based cell architectures including DMFCs and redox flow batteries, which demonstrates good potential for low-power commercial applications.

Another notable result from the comparison in Table 6.1 is the remarkably high performance achieved with the recent microfluidic cells developed by Da Mota et al. and Braff et al. First, the co-laminar borohydride fuel cell by Da Mota et al. uniquely exploited the favorable combination of a porous convective flow barrier (i.e., porous separator) and half-cell confined chaotic mixing induced by herringbone groove-patterned electrodes [20]. By direct comparison against a non-patterned baseline, the chaotic flow was shown to double the power density from ~125 to 250 mW cm^{-2}. Due to the use of considerably higher flow rates, different reactants with a higher cell potential, mixed media conditions (alkaline anode and acidic cathode), a nitric acid electrolyte that may also be consumed during reaction, and the presence of catalysts, a direct comparison with the other cells presented in Table 6.1 is dubious. At least half of the gain in volumetric power density can be attributed to the thinner aspect ratio of the cell with a reduced distance between the electrodes. Due to the reduced chamber volume, higher flow rates, and spiraling secondary flow, it is likely that the parasitic pumping power was significantly higher than for the other cells. Nevertheless, it was reported that the cell compares favorably with high-performance conventional membrane-based hydrogen fuel cells, which is a considerable achievement. Second, the hydrogen/bromine microfluidic fuel cell developed by Braff et al. [21] also pushed the envelope for performance by a considerable margin. In this case, the air-breathing design first proposed by Jayashree et al. [22, 23] was strategically modified by switching the gas diffusion electrode to the hydrogen anode known have very high electrochemical activity on Pt and replacing the relatively slow oxygen cathode with a kinetically favorable redox couple in the form of bromine/hydrobromic acid. With roughly the same architecture as the original cell, the performance benefits can likely be attributed to the improved reactant chemistries and reaction rates as well as the enhanced transport properties provided by the small gaseous hydrogen molecules and the relatively high concentration of bromine. More detailed information and experimental results would however be required to adequately assess this novel cell configuration. In either case, the outstanding performance further underlines the benefits of the microfluidic fuel cell technology as a precision manufacturing method for on-chip power applications.

Last, the cell by Zebda et al. [24] highlights the importance of microfluidic cell architectures in the biofuel cell domain. As the highest efficiency biofuel cell reported at the time of publication, it featured a Y-channel design similar to the original cell by Ferrigno et al. Notably, the performance of biofuel cells is predominantly limited by the kinetics of the enzymatically catalyzed reactions which are orders of magnitude lower than for nonbiological electrocatalysts [25]. More importantly, this contribution provides a useful benchmark that demonstrates the value of microfluidic cells as analytical platforms for testing other technologies such as bioelectrodes. Biofuel cells and biobatteries stand to benefit from development of microfluidic cell technologies through (1) enhancing convective mass transport, enabling higher enzyme loading while maintaining the enzymatic turnover rates close to their full capacity; (2) harnessing the high surface-to-volume ratio inherent to microstructured devices to promote the surface-based electrochemical reactions catalyzed by the immobilized enzymes; and (3) providing useful scale-up opportunities for practical devices with automated reactant supply and on-chip integration compatible with MEMS technologies. Moreover, microfluidic devices with automated reactant supply are well suited for stability analyses as they provide steady state operation under fixed conditions while keeping each enzyme in an optimum environment.

6.3 Utility

Besides performance comparisons based on power, secondary performance metrics such as fuel utilization, voltage efficiency, and coulombic efficiency are also useful to understand the sources of performance losses but are less commonly available in the literature. Durability is another supplementary metric expected to become increasingly important as the technology matures, especially for biofuel cells. As previously mentioned, architectural alterations are most likely to be tailored towards performance enhancements, whereas fabrication methods are likely to address issues pertaining to cost, device durability, and system integration. In terms of overall utility, microfluidic electrochemical cell technology is likely to evolve according to the following two application-oriented subgroups: planar, monolithic devices are well suited to be integrated directly into on-chip applications, while multilayer devices are poised to become stacked for medium-power (~1–10 W) applications.

Most of the pros and cons of the microfluidic cell technology are known only qualitatively at this stage. The primary advantages are known to be reduced cost and longer durability due to the absence of a membrane, electrolyte pH flexibility, and the compatibility with single-layer on-chip manufacturing. Another important benefit is the fabrication flexibility of this technology which is amenable to laser etching, CNC machining, lithography, conventional multilayer MEA fabrication, and possibly even 3-D printing. The disadvantages on the other hand are likely to be related to increased ohmic losses due to wider electrode separation, increased sensitivity to pressure fluctuations which could disrupt the co-laminar interface, lower recirculation possibilities due to reactant mixing, and considerable engineering challenges with scale-up or stacking of devices. Once these issues have been more definitively assessed, a cost-benefit analysis could reveal whether the microfluidic cell technology is economically viable for specific applications.

According to the old architectural adage form follows function, the appropriate design of microfluidic cells must satisfy certain functional requirements. Occasionally, certain forms fortuitously result in unintended functions coming to light. With colorful reactants such as vanadium ions, the original planar cell design in transparent substrate material allows for excellent visualization of both reduction and oxidation reactions. This attractive feature enables convenient diagnosis of manufacturing quality and cell performance issues, but can also be exploited in a much broader context. Combined with the relative ease of fabrication, the unique visualization capabilities have allowed microfluidic cells to be utilized as instructional tools to engage students in the classroom [26]. Given the advent of 3-D printing with continuously improving dot resolutions [27], microfluidic devices could become even more widespread, making this technology readily accessible to research groups with limited funding who may apply standardized units as a test bed for other electrochemical cell components such as specialized electrodes using immobilized biocatalysts [28] or for detection of microorganism electroactivity [29]. The increasing number of reports such as these indicates that research concerning microfluidic electrochemical cells is both expanding and bifurcating into two areas: studies which attempt to modify and improve existing cell designs and those which use microfluidic cells as analytical platforms to investigate other components. In both cases, lessons can be learned with microfluidic cells that may have repercussions for other technologies even beyond the general field of electrochemical energy conversion and storage. In sum, microfluidic electrochemical cells may come to serve equally important functions as analytical research tools in addition to commercial and educational efforts.

Overall, the performance of microfluidic cells in terms of power density has already reached or even exceeded the levels of comparable technologies. Performance benchmarks aside, other critical metrics related to utility also need to be addressed in order for this technology to become commercially viable, namely the efficiency/fuel utilization, co-laminar interface stability, and the stacking/scale-up solutions. Although nearly 100 % fuel utilization has been achieved [16], matching this level of efficiency with high power density is a major challenge that has not been realized to date. Further research on recirculation may potentially lead to a practical solution in this regard, subject to adequate crossover protection schemes being developed, while additional fundamental work on material development and optimization would also be required. Regarding interface stability, few studies have been performed on the effects of pressure fluctuations and dynamic operations. Due to the small distances between electrodes and microliter chamber volumes, it is unlikely that co-laminar cells could withstand pressure fluctuations without the associated performance fluctuations, obviating the need for more quantitative work on this challenge. In addition, there is a lack of engineering solutions for important system functions including the integration of fuel and oxidant storage, waste handling, and low-power microfluidics based fluid delivery using integrated micropumps and microvalves. With integrated infrastructure, microfluidic fuel cell and battery modules could be integrated on-chip as independent power sources for various MEMS devices and microfluidic systems, as stand-alone units or in hybrid configurations in combination with small secondary batteries or capacitors. In terms of capacity increase of single cells, there are physical size limits outside of which co-laminar flow is not practical (cf., Chapter 2) and scale-up of individual cells is unlikely without the added use of separators [30]. The challenge associated with cell expansion is perhaps the most relevant technological issue to be addressed by research in this field, and a viable solution would require improved fundamental understanding of two key elements, namely (1) the maximum cell dimensions and reactant flow rates amenable to reliable co-laminar flow and (2) the maximum power levels of expanded single cells and whether these levels can match the requirements of commercial applications. Establishing reproducible protocols to quantify these points would elucidate the maximum size and power for cells relying strictly on co-laminar flow versus the added use of porous separators and would strategically address application-oriented technology matching and development. Parametric studies such as the one by Thorson et al. [31] would need to be replicated and expanded with a combined modeling and experimental approach in the context of the design principles outlined in Chapter 2 of this book. Scale-up of flow cell systems through stacking and/or multiplexing is another possible alternative to single-cell scale-up. To date, published work involving multiple cells has primarily considered monolithic planar configurations [32–34], thereby achieving a proportionally higher power output at the expense of a considerably larger microfluidic chip. Sandwich structure fabrication, preferably with multilayer cell designs, is more amenable to stacking due to the long history of development for conventional MEA-based fuel cell stacks and would therefore be the next essential step. Regardless of fabrication method, however, it is important to investigate whether maintaining balanced electrolyte pressures will become increasingly challenging for multiple cells at a time.

While the proposed research directions outlined above are deliberated, it may also be useful to consider a change of perspective. Perhaps the true challenge for microfluidic fuel cells and batteries is the question of appropriate application matching and system integration. Finding a stable or stationary low-power commercial application to drive research and technology development in this field would likely require minimal stacking and no special pressure control. One of the most promising applications identified to date may be the wireless sensor networks used in remote or off-grid locations which would benefit considerably from the relatively low cost and minimal maintenance of microfluidic power sources [35]. Another promising application matching opportunity for microfluidic electrochemical cells is to provide "two-in-one" chip-integrated power and cooling of microelectronics. In this case, microfluidic solutions could deliver high power density and liquid cooling directly to the most critical sites of high power consumption and associated heat generation. This technology opportunity is currently investigated by IBM and Ecole Polytechnique Fédérale de Lausanne (EPFL) for high-performance computer architectures, an application for which microfluidic electrochemical cells may have a beneficially disruptive effect [36]. In order to reach the functional prototype demonstration phase for these applications, however, more research needs to be conducted across the entire microfluidic system which would include reservoirs, manifolds, and a microscale pumping mechanism for the reactants. Such a wide scope would benefit from active collaboration within the microfluidics community and could be accelerated through strategic university–industry R&D partnerships.

References

1. F.R. Brushett, H.T. Duong, J.W. Ng, R.L. Behrens, A. Wieckowski, P.J.A. Kenis, J. Electrochem. Soc. **157**, B837–B845 (2010)
2. F.R. Brushett, M.S. Naughton, L. Ng, J.W.D. Yin, P.J.A. Kenis, Int. J. Hydrogen Energ. **37**, 2559–2570 (2012)
3. H.-R. Jhong, F.R. Brushett, L. Yin, D.M. Stevenson, P.J.A. Kenis, J. Electrochem. Soc. **159**, B292–B298 (2012)
4. M.S. Naughton, F.R. Brushett, P.J.A. Kenis, J. Power. Sources **196**, 1762–1768 (2011)
5. C.M. Moore, S.D. Minteer, R.S. Martin, Lab Chip **5**, 218–225 (2005)
6. S.A. Mousavi Shaegh, N.-T. Nguyen, S.M. Mousavi Ehteshami, S.H. Chan, Energ. Environ. Sci. **5**, 8225–8228 (2012)
7. A.K. Shukla, R.K. Raman, K. Scott, Fuel Cell. **5**, 436–447 (2005)
8. W. Sung, J.-W. Choi, J. Power. Sources **172**, 198–208 (2007)
9. M. Togo, A. Takamura, T. Asai, H. Kaji, M. Nishizawa, Electrochim. Acta **52**, 4669–4674 (2007)
10. M. Togo, A. Takamura, T. Asai, H. Kaji, M. Nishizawa, J. Power. Sources **178**, 53–58 (2008)
11. S. Tominaka, S. Ohta, H. Obata, T. Momma, T. Osaka, J. Am. Chem. Soc. **130**, 10456–10457 (2008)
12. S. Topcagic, S.D. Minteer, Electrochim. Acta **51**, 2168–2172 (2006)
13. J. Larminie, A. Dicks, *Fuel Cell Systems Explained*, 2nd edn. (Wiley, Chichester, 2003)
14. S.A. Mousavi Shaegh, N.-T. Nguyen, S.H. Chan, W. Zhou, Int. J. Hydrogen Energ. **37**, 3466–3476 (2012)
15. J.W. Lee, E. Kjeang, J. Power. Sources **242**, 472–477 (2013)
16. E. Kjeang, R. Michel, D.A. Harrington, N. Djilali, D. Sinton, J. Am. Chem. Soc. **130**, 4000–4006 (2008)
17. M. Skyllas-Kazacos, M.H. Chakrabarti, S.A. Hajimolana, F.S. Mjalli, M. Saleem, J. Electrochem. Soc. **158**, R55–R79 (2011)
18. P. Zhao, H. Zhang, H. Zhou, J. Chen, S. Gao, B. Yi, J. Power. Sources **162**, 1416–1420 (2006)
19. C. Xu, A. Faghri, X. Li, T. Ward, Int. J. Hydrogen Energ. **35**, 1769–1777 (2010)
20. N. Da Mota, D.A. Finkelstein, J.D. Kirtland, C.A. Rodriguez, A.D. Stroock, H.D. Abruña, J. Am. Chem. Soc. **134**, 6076–6079 (2012)
21. W.A. Braff et al., Nat. Commun. **4**, 2346 (2013)

22. R.S. Jayashree, L. Gancs, E.R. Choban, A. Primak, D. Natarajan, L.J. Markoski, P.J.A. Kenis, J. Am. Chem. Soc. **127**, 16758–16759 (2005)
23. R.S. Jayashree, S.K. Yoon, F.R. Brushett, P.O. Lopez-Montesinos, D. Natarajan, L.J. Markoski, P.J.A. Kenis, J. Power. Sources **195**, 3569–3578 (2010)
24. A. Zebda, L. Renaud, M. Cretin, C. Innocent, R. Ferrigno, S. Tingry, Sens. Actuators B **149**, 44–50 (2010)
25. F. Gao, L. Viry, M. Maugey, P. Poulin, N. Mano, Nat. Comm. **1**, 1–7 (2010)
26. K. Davis, J. Muskin, Mater. Res. Soc. Symp. Proc. **1320**, (2011)
27. T. Serra, J.A. Planell, M. Navarro, Acta Biomater. **9**, 5521–5530 (2012)
28. Z. Li, Y. Zhang, P.R. LeDuc, K.B. Gregory, Biotechnol. Bioeng. **108**, 2061–2069 (2011)
29. H.-Y. Wang, J.-Y. Su, Bioresour. Technol. **145**, 1–4 (2013)
30. E. Kjeang, N. Djilali, D. Sinton, J. Power. Sources **186**, 353–369 (2009)
31. M.R. Thorson, F.R. Brushett, C.J. Timberg, P.J.A. Kenis, J. Power. Sources **218**, 28–33 (2012)
32. R. Ferrigno, A.D. Stroock, T.D. Clark, M. Mayer, G.M. Whitesides, J. Am. Chem. Soc. **124**, 12930–12931 (2002)
33. B. Ho, E. Kjeang, J. Fluid Eng. **135**, 021304 (2013)
34. K.S. Salloum, J.D. Posner, J. Power. Sources **196**, 1229–1234 (2011)
35. W. Gellett, M. Kesmez, J. Schumacher, N. Akers, S.D. Minteer, Electroanalysis **22**, 727–731 (2010)
36. M.M. Sabry, A. Sridhar, D. Atienza, P. Ruch, B. Michel, in *Proceedings of the IEEE/ACM 2014 Design Automation Test Europe conference*, 2014, pp. 70–75
37. M.A. Goulet, E. Kjeang, Co-laminar flow cells for electrochemical energy conversion. J. Power. Sources **260**, 186–196 (2014)

Chapter 7
Conclusions and Recommendations

Considering that the invention of the microfluidic fuel cell is relatively recent, the number of advances in this field is impressive. Devices have been developed based on various fuels and oxidants, with competitive power densities and cell voltages obtained at room temperature. The levels of fuel utilization have been raised from below 1 % to nearly 100 % per single pass in some cases. Many of these advances, as discussed in this book, have stemmed from improving transport through microfluidic fuel cell architecture and running conditions. Several scale-up methodologies have also been demonstrated that show promise in translating advances made in unit cells to highly functional integrated devices. Based on current publication and citation trends, significant further growth is anticipated in this field.

Among recent developments, co-laminar flow cell architectures utilizing flow-through porous electrodes were shown to achieve power densities surpassing those of previous flow-by designs and even conventional electrochemical flow cell technologies. This enhancement has expanded the functionality of the co-laminar flow-based technology to include rechargeable battery operation. Porous separators were recently demonstrated to further mitigate reactant crossover and stabilize the co-laminar flow, enabling patterned electrodes to be used for increasing mass transport through convection. The air-breathing cathode was extended to a hydrogen breathing anode with record breaking volumetric power density in the context of membraneless microfluidic cells. In regards to cell design and utility, two important trends were identified, namely multilayer sandwich structures showing potential for stacked operation and medium-power applications and single-layer monolithic construction for low-power on-chip applications.

Although the initial development of prototype microfluidic fuel cell and battery devices has been rapid, much further work is required to facilitate a major commercial breakthrough. In this context, application driven research and technology development is essential, an approach that has been critical to the success of many other power source technologies. Ideally for pairing with consumer electronics, the microfluidic fuel cell system, including auxiliary equipment and fluid storage, would provide a power output in the 1–20 W range in a compact integrated package with simple connections to established external infrastructure. This is a tall order, most amenable

E. Kjeang, *Microfluidic Fuel Cells and Batteries*, SpringerBriefs in Energy,
DOI 10.1007/978-3-319-06346-1_7, © The Author(s) - SpringerBriefs 2014

to the multilayer cell architecture with its associated stacking opportunities. Alternatively, a low-cost microfluidic power system could be developed for wireless sensor networks and other low-power chip-integrated devices, based on the existing monolithic on-chip cell architecture supported by a simple fluid handling system.

The power density comparisons made in this book indicate that the performance of membraneless microfluidic fuel cells and batteries has already reached or even exceeded the levels of similar MEA-based cells, including vanadium redox flow batteries and direct methanol fuel cells. Further increases in power density can be anticipated as the technology matures. In the biofuel cell and biobattery domain, the ongoing research on specialized microfluidic cell architectures may lead to a critically sought paradigm shift on performance. Volumetric normalization of power density and standardized testing conditions are generally recommended to benchmark new devices. At the present stage of development, however, the technology would particularly benefit from targeted research towards secondary performance metrics such as fuel utilization, voltage efficiency, and coulombic efficiency—all of which are equally critical for device integration. Comprehensive research efforts must also address the challenges and opportunities associated with pressure fluctuations, reactant recirculation, in situ recharging, and cell expansion and stacking methods. Engineering solutions for system aspects such as integration of reactant storage, waste handling, and low-power fluid delivery using integrated micropumps and microvalves are likewise required. Where possible, existing solutions developed for other electrochemical cells or microfluidic devices could be adapted to microfluidic cells. With definitive advances in these areas, a cost-benefit analysis could help guide the development towards economically viable applications.

The emerging use of microfluidic fuel cells and batteries for analytical applications and educational purposes is also encouraging. The low cost, fabrication flexibility, and unique visualization capabilities inherent to microfluidic cells make them well suited as instructional tools to engage students in the classroom, potentially for a wide variety of courses in the areas of energy conversion and storage, applied chemistry, and microsystems. For analytical applications, standardized units could be produced as a convenient, low-cost platform for in situ lab-scale testing and characterization of electrochemical cell components such as novel electrocatalysts, catalyst supports, and bioelectrodes. Overall, microfluidic electrochemical cells may come to serve equally important functions as analytical and educational tools in addition to commercial utility.

About the Author

Dr. Erik Kjeang is an Assistant Professor in Mechatronic Systems Engineering and Director of the Fuel Cell Research Laboratory (FCReL; http://www.fcrel.ca) at Simon Fraser University (SFU) in Vancouver, Canada. Dr. Kjeang holds a Ph.D. in Mechanical Engineering from the University of Victoria (UVic), Canada, and an M.Sc. in Energy Engineering from Umea University, Sweden. His research program at SFU encompasses the general area of sustainable energy technologies with specialization in electrochemical power generation. Prior to joining SFU, Dr. Kjeang worked as a research engineer at Ballard Power Systems, a world leader in hydrogen PEM fuel cell development and manufacturing. He is an established expert in fuel cell science and technology and has authored more than 100 peer-reviewed publications, developed patented technology, and given several invited lectures at major international conferences in this field. His feature research on microfluidic fuel cell technology was awarded with the prestigious Governor General's Gold Medal for outstanding dissertation and numerous other awards and fellowships. Dr. Kjeang is currently the Principal Investigator of a $12M Automotive Partnership Canada-supported university–industry collaborative research project on heavy duty bus fuel cells involving Ballard, SFU, and UVic. The fundamental understanding and new technologies developed by his team has improved the performance and durability of Ballard's fuel cell systems and reduced the manufacturing cost.

E. Kjeang, *Microfluidic Fuel Cells and Batteries*, SpringerBriefs in Energy,
DOI 10.1007/978-3-319-06346-1, © The Author(s) - SpringerBriefs 2014

Index